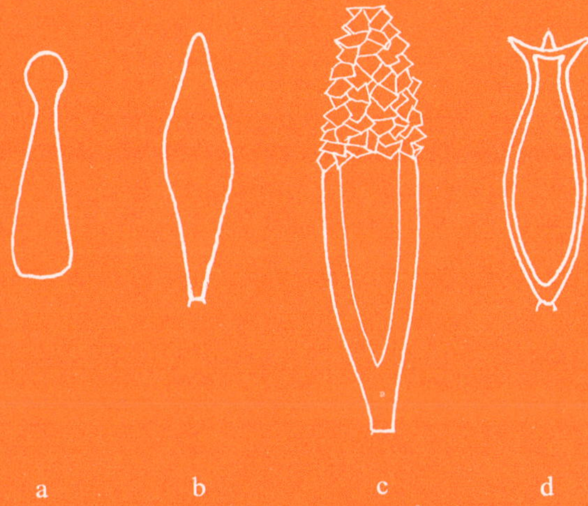

Examples of cystidea shapes

(a) capitate (b) spindle shaped (c) metuloid with crysta

(d) metuloid with projections

Microscopic characters of selected basidiomycete fungi illustrating basidia, cystidea, and spores.

(a) *Phylloporus clelandii* (b) *Russula clelandii* (c) *Inocybe* sp.

(d) *Entoloma viridomarginatum*

Dermocybe splendida

Cystoderma amianthinum

A field guide to AUSTRALIAN FUNGI

Bruce Fuhrer

Including a range of name changes since the 2005 first edition.

First published in Australia in 2005 by
Bloomings Books Pty Ltd
Reprinted with revisions, 2009
Reprinted with additional revisions in 2016
Reprinted in 2020, 2022 and 2024
Melbourne, Australia
www.bloomings.com.au

Bloomings Books Pty Ltd is a specialist publisher and distributor of natural history and horticultural books.
Phone: +61 (0)4 3818 2801

The National Library of Australia Cataloguing-in-Publication entry:

Fuhrer, B. A. (Bruce Alexander), 1930- .
A field guide to Australian fungi.

Bibliography.
Includes index.
ISBN 1 876473 51 7.

1. Fungi - Australia - Identification. I. Title.

579.50994

Book Designers: Ranya Langenfelds, Luke Harris
Editor: Helen Young
Photography: Bruce Fuhrer
Publisher: Warwick Forge
Printed in China by Everbest Printing Investment Ltd.

Contents

Acknowledgments

I am most grateful for the encouragement and assistance of Margaret Corrick. She has worked closely with me in the preparation of the text and also compiled the glossary and index. Her help was invaluable in bringing the project to fruition.

Cheryl Grgurinovic of Canberra edited a preliminary draft of the manuscript and provided taxonomic assistance, particularly in the genus Mycena.

Tony Young of Queensland assisted with Hygrocybe taxonomy, and Teresa Lebel of the National Herbarium of Victoria provided taxonomic assistance and helpful comments on the manuscript.

My close friends, George Crichton, the late Gordon Beaton, the late Cliff Beauglehole and the late Jim Willis shared their interest in fungi over many decades. I will always be grateful for their inspiration and encouragement.

I shared many fungal forays with my dear friend, the late Ian McCann, who kindly provided the photograph of *Phallus indusiatus*.

My wife Irene, daughter Robyn, and grandchildren Christopher, Ashleigh and Lachlan have been a constant source of encouragement, and helpful companions in the field.

Bruce Fuhrer
Ringwood North
October 2004

Coprinus comatus

Introduction

The fungal kingdom has long been a source of fascination, mystery and even of fear. The rapid appearance after rain of the colourful or strangely formed fruiting bodies of the large-fruited (macrofungi) seems almost magical. Their beauty is a continual inspiration to artists and photographers. For centuries myths and fairy tales have been woven around them and many are valued as human and animal food. In recent years their importance in the ecology of the planet has been more widely appreciated and many individuals and amateur groups have become involved in the study of the larger and more conspicuous species.

This book aims to encourage interest in the study of these fungi and to assist in the identification of fruit bodies. Many of the species illustrated are widely dispersed in Australia and some also occur overseas. Habitat notes are included in each caption but geographical distribution is too poorly known to make its inclusion meaningful. The increasing use of wood chips as garden mulch has led to the appearance of many interesting fungi in suburban gardens and parks. Some species are associated with particular habitat types or plant associations and several exotic species have spread because of their association with particular trees.

Fruiting times are not stated as these are dependent on seasonal conditions. In southern Australia, peak fruiting will usually occur from May to July, if good rain has fallen while the soil is still warm. Tropical species fruit early in the wet season.

Nomenclature follows the Interactive Catalogue of Australian Fungi available on the web site of Royal Botanic Gardens, Melbourne. Where synonyms or alternative names are given, the currently accepted name is printed in bold.

It has not been possible to give species names to all the fungi illustrated in the book as a large proportion of Australian fungi are at present unnamed. However they are included because many are relatively common or of particular interest.

Only a few well-known and appropriate common names in use in Australia have been given. The application of European vernacular names to our fungi is not encouraged and can be misleading, particularly if names applied to edible European species are used for superficially similar Australian species which are inedible or potentially toxic.

Biology of the macrofungi

Fungi differ from plants in that they do not contain chlorophyll and are thus unable to obtain nourishment by photosynthesis. They must therefore rely on the absorption of nourishment from the substrate. This is done by colonization of their chosen substrate with thread-like filaments called **hyphae**. These form a web-like mass called the **mycelium** which spreads through the substrate, releasing enzymes and absorbing nutrients. The mycelium is rarely seen but occasionally bundles of hyphae will form thick threads or cords called **rhizomorphs**, as in *Stereum illudens* (**441**).

On open forest floors and in grassy places fungi often appear in rings. The mycelium of the fungus expands outwards in its search for food and the fruit bodies appear each season in expanding rings, often known as fairy rings. Sometimes the growth of grass in front of the mycelium is stimulated and then suppressed behind the mycelium, giving rise to circles of suppressed or vigorous growth resulting in multiple rings.

In simple mushrooms the gills are exposed, but in some genera the young gills are covered by a partial veil, which can be membranous, as in *Amanita*, or by a cortina, a cobweb-like veil, as in *Cortinarius*. The partial veil ruptures as the fruit body expands, leaving either an annulus (or ring) on the stem or sometimes fragments hanging from the margin of the cap, or fibrils adhering to the stem.

In some genera a membranous universal veil totally envelopes the button stage of the fruit body. The veil ruptures as the stem elongates and the cap expands, leaving scaly or mealy remnants on the cap surface, and a cup-like sheath or torn rim (volva) at the base of the stem. In other genera

the universal veil can be membranous, mealy or just a slimy coating.

Some groups of fungi have a modified structure that suits their habitat requirements. Species of *Crepidotus* and *Panellus* have a reduced lateral stem and form gilled brackets on wood. The boletes and polypores have a layer of pores instead of gills, with the fertile surface lining the inside of the pores. The hydnums and relatives have spines covered with fertile tissue. The coral fungi and jelly fungi have the fertile surface covering the fruit body above the stem, if present.

The macrofungi in nature

Fungi are an essential part of the ecosystem. The macrofungi have several important roles in plant ecology, such as the formation of mycorrhizal associations and saprophytic and parasitic relationships.

The majority of plants, including trees, usually form a symbiotic relationship with a number of fungi and together form specialized roots known as **mycorrhiza**. This association provides the fungi with sugars and water and assists the plant with mineral and nutrient uptake. It also confers antibiotic protection for the plant against pathogenic fungi and other soil-borne organisms.

In contrast to the mycorrhizal fungi that assist plant growth, the **saprophytic** fungi are the major agents in the breakdown and recycling of dead plant and animal material and of all types of organic waste materials. These fungi are common in the forest on dead trees and wood, leaf litter and animal dung. They will also be found on decaying household rubbish. Some saprophytes are capable of weakening or destroying structural timbers in buildings, outdoor furniture and fences, especially in damp places and tropical climates. Old or unhealthy trees frequently have their heartwood consumed by fungi, a condition known as 'heart rot'. Although the outer sapwood remains intact the tree is structurally weakened and the value of the timber is reduced.

Some of the large-fruited fungi are plant **parasites** causing disease and death of cultivated and native species. *Armillaria* species are probably the best known of these. The fungus spreads from infected plants by creeping, underground, string-like rhizomorphs that invade the roots of neighbouring plants. In healthy forests the plants are well equipped to repel pathogenic fungi and it is usually the old, stressed or damaged plants that become infected.

Specialized fungi also parasitize insects and other arthropods, including spiders. The most familiar are species of *Cordyceps* that parasitize underground larvae of moths and beetles. The larval tissue is replaced by the fungal tissue which produces a club-shaped fruit body. Sometimes dead spiders sprouting fungal fruit bodies are found on shaded rock faces. On forest tree trunks the mummified remains of hairy caterpillars are often seen with dark hairs emerging through the white fungal tissue.

Fungi as food

Fungi form a major component of the diet of some of our native animals. Some small wallabies and potoroos can detect and excavate underground truffle-like species. Other small animals such as bandicoots and wombats and some birds also include fungi in their diets. Several fly and beetle species deposit their eggs on or in fungi, which become food for the larvae. Springtails, mites, snails and slugs also feed on fungi.

Some fungi are used for their medicinal properties. The hard, woody bracket fungus *Ganoderma lucidum* is widely cultivated in Asia, and some *Cordyceps* species are valued in Asia and Africa.

Fungi have been highly prized as human food since historic times. Charles Darwin found a tribe in Tierra del Fuego that subsisted on fresh or cured mushrooms, supplemented with what little meat they could find. In some cold areas of northern Russia pickled or dried fungi are the main winter vegetable. Well-known edible species in Australia include the field mushroom *Agaricus campestris* and a few other *Agaricus* species such as *A. bisporus*, the white cultivated mushroom of commerce. Other edible species, all associated with pine trees, are the Saffron Milk Cap (*Lactarius deliciosus*), Slippery Jack (*Suillus luteus*) and the related *S. granulatus*. Many fungus species are now cultivated commercially and are sold fresh, and as many as fifty species are imported dried. The most sought after and expensive edible fungus is a European truffle, native to the oak forests of northern Europe, and now being cultivated in Australia.

This book is not intended as a guide to the edibility of wild fungi in Australia and it should be remembered that collection of fungi from public land in most parts of Australia is illegal without a permit.

Poisonous fungi

Because of the difficulty of accurately identifying edible species, apart from the common field mushroom and a few well-known pine forest species already mentioned and one or two others noted in the text, all fungi should be regarded with suspicion. There is no reliable rule for recognizing edible species, apart from experience or having the specimen expertly identified. Several particularly toxic species are included in the book and are noted as such in the captions. Quantities as small as a gram or two of *Amanita phalloides*, the Death Cap, usually found under mature oak trees, can be fatal, or cause permanent liver and kidney damage.

Distribution of macrofungi

Most fungus species have a much wider distribution than flowering plant species. Many of the same fungus species occur throughout the temperate near-coastal regions of Australia. The tropical, temperate and cool temperate rainforests each have their own fungal association, which is mostly restricted to these habitats or to similar microhabitats elsewhere.

Arid and semi-arid habitats have a distinctive fungus community usually dominated by puffballs and their relatives, with other less common species such as the peculiar black *Montagnea arenaria* and the tall, stalked *Podaxis beringamensis* which grows on termite mounds. Fruiting of fungi in arid regions is sporadic but after soaking rains they may fruit freely.

Some fungi are associated with specific plants or plant genera. *Cyttaria gunnii* and several species of *Cortinarius* are associated with *Nothofagus* species. *Underwoodia beatonii* is usually found with *Melaleuca* species. *Banksiamyces* are specific to *Banksia* cones. Some of the underground truffle-like fungi have specific host-plant and animal associations that affect their distribution. The species that are eaten by animals are dispersed by the spores that are passed out in the faeces.

Fires stimulate mass fungal fruiting in some species such as *Laccocephalum mylittae* and *L. tumulosum*, which have underground storage organs known as sclerotia or pseudosclerotia. The mineral-rich conditions induced by fire appear to encourage fruiting of *Morchella elata* and *Anthracobia muelleri*, while *Pholiota highlandensis* usually fruits on charcoal beds.

Classification of fungi

When the scientific classification of the world's living organisms started in the middle of the 18th century it was considered that all living things could be separated into two kingdoms, plants and animals. Organisms that moved were placed in the Animal Kingdom, those that didn't were considered to belong to the Plant Kingdom. Although the fungi are clearly very different from plants in that they do not obtain their food through photosynthesis, the two-kingdom classification was retained until the second half of the 20th century. The development of the electron microscope, studies in fungal biochemistry and, more recently, studies of DNA have led to the realisation that more than one kingdom would be needed to encompass the diversity of the fungi.

Australian authorities recognize three kingdoms, Protoctista, Chromista and Eumycota. A detailed discussion of fungal classification is beyond the scope of this book. Interested readers are referred to the accompanying list of additional reading, particularly the *Fungi of Australia* series and the web sites of the Australian Biological Resources Study and of the Royal Botanic Gardens, Melbourne.

Except for the illustrations of slime moulds, which belong to the kingdom Protoctista, all the fungi dealt with in this book belong to two divisions of the kingdom Eumycota, often referred to as the 'true fungi'. All the large-fruited forms commonly known as mushrooms and toadstools belong in these divisions, as well as a great variety of other species with conspicuous fruit bodies.

The **Basidiomycota** (Section 1) produce their spores on microscopic club-shaped cells called basidia. Each basidium usually has four spores, but sometimes only two. The basidia are arranged on gills, in pores or on the outer surface of the fruit body, depending on the genus.

Fig. 1 Basidium with spores

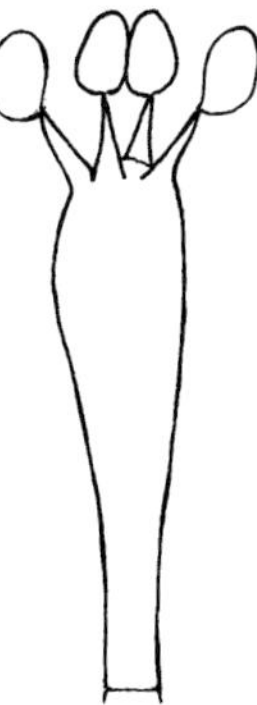

Ascomycota (Section 2), produce their spores inside microscopic, elongated sac-like cells called asci. Each ascus usually produces eight ascospores. When mature, the ascus ruptures at the top and the spores are shot out into the air. The asci are usually formed on the fertile surface of the fruit body but sometimes inside a small flask-shaped structure called a perithecium that is embedded in the flesh of the fruit body. When the spores are mature, the perithecium opens through a minute pore at the top called the ostiole. The Ascomycetes include cup fungi and relatives, vegetable caterpillars and flask-fungi.

Fig. 2 Ascus with spores

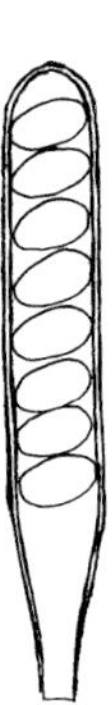

Myxomycota (Section 3) covers slime moulds, which are not true fungi. These organisms belong to the kingdom Protoctista, division Myxomycota. Illustrations of a few species are included because of the interest usually aroused by the often strikingly coloured, mobile plasmodial stage. At maturity the mobile stage condenses into hard, spore-producing organs.

Spore Prints

Spore characters are important aids to identification. Spore colour is particularly useful and can be easily determined by making a spore print. This is best done by placing a mature fruiting body, fertile surface down, on a sheet of white paper away from draughts. Leave the specimen for several hours, or overnight, by which time the spore print should be clearly seen. If the specimen is suspected of having white spores, dark paper should be used. Size and shape of individual spores is critical in accurate determination of species, but can only be seen with the aid of a microscope. Drawings of a selection of individual spores from different species are shown on the endpapers of this book.

Fig. 3 Spore prints

Recorder's name B & I. Fuhrer **Reference No** B850 **Date** 10/5/89

Location Narbethong Pinus r. plantation. Anderson's Rd. Narbethong

Name of fungus if known: Genus *Amanita* Species *muscaria*

Cap description (sketch if possible)

Size range
Shape range
Texture
Colour
Veil present or not

70-210 mm. Convex when young, covered with white warts, older caps flattened, smooth, warts mos washed off in older caps. Brilliant crimson. Universa veil covers button stage.

Gill description (sketch if helpful)

Spacing
Shape
Nature of edge
Colour range
Attachment to stem

Gills very close, broad, straight to ventrally curved, pale cream. Free from stem. Covered by partial veil when immature.

Stem description

Length
Diameter
Shape including base
Texture
Colour
Ring or not

Up to 150 mm high, 20 mm diameter, parallel with bulbous base with rings of warty tissue. Large ring around upper part of stem.

Spore print colour
Spore deposit if possible

White

Odour None, faint

Habitat description and growth habit of fungus

Growing in scattered troops on ground in *Pinus radiata* plantat
Locally common, associated with Suillus luteus.

Other observations

Recording field characters and data

As fungi decay quickly it is important to make notes in the field. Field descriptions should follow the example on the opposite page.

Arrangement of illustrations

The two main sections of the book, dealing with the divisions Basidiomycota and Ascomycota, are each further divided into informal groups based mainly on the gross morphology of the fruit bodies rather than on true taxonomic relationships. Within each group the arrangement is alphabetical and, when used in conjunction with the illustrated guide to groups, the reader should easily be able to search for similar specimens to the one in hand. Every effort has been made to keep up with changes in nomenclature.

In the caption headings current names appear in ***bold italics***, with synonyms and previously accepted names bracketed and in *light italic* type. Due to the design challenge presented by the presence of so many illustrations in a small book, some captions have had to be inserted overleaf from their pictures. In each case arrows in the picture and beside the caption (→) indicate that this has occurred. The photographs are not taken to a common scale but appropriate measurements are stated in each caption.

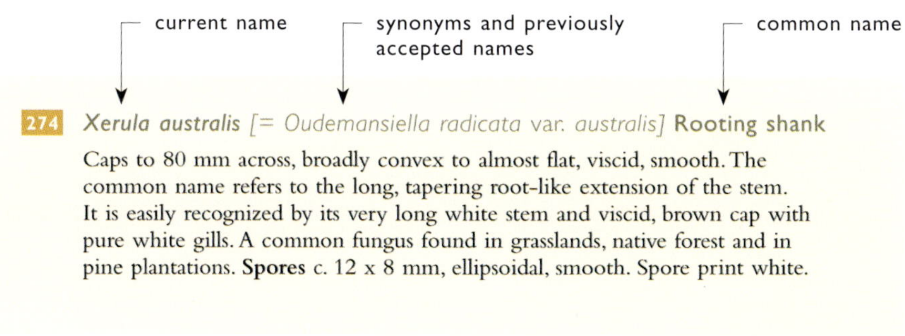

Note regarding 2016 edition: Species continue to be listed as previously published but with the old name first and the new species following inside square brackets. All names changed in the 2016 edition are marked with an asterisk. All recent name changes are also listed in the schedule on page 356.

Recent name changes

Since the first publication of this book in 2005 there have been a number of name changes and these are now included in this publication. Species continue to be listed as previously published but with the former name first and the new species following within squared brackets. All recent name changes are also listed in the Schedule of Recent Name Changes on p355. Both old and new names will also be found in the Index.

PICTORIAL Guide to Groups

Section 1

BASIDIOMYCOTA

Group	Fertile surface	Texture	Page
Agaricus & allies (Fungi with gills) 1.1	Simple gills	Fleshy	18
Paxillus & allies (Fungi with forked gills) 1.2	Forked gills	Fleshy	181
Boletus and allies (Fleshy Pore-fungi) 1.3	Pores	Fleshy	186
Coral & Club fungi 1.4	Exterior of fruit bodies *Clavaria* – spores white *Ramaria* – spores yellow-brown	Fleshy	196

Group	Fertile surface	Texture	Page
 Soft puffballs 1.5.1	Internal	Firm to soft	213
 Earth Stars 1.5.2	Internal	Firm to soft	217
 Hard-skinned puffballs 1.5.3	Internal	Hard to leathery	217
 Bird's-nest fungi 1.5.4	Contained in peridioles	Soft to firm	221
 Stalked puffballs 1.5.5	Terminal spore mass	Firm to hard	223
 Stinkhorns 1.5.6	Slimy gleba	Fleshy	228
 Underground or truffle-like fungi 1.6	Internal	Fleshy	237
 Spine fungi (*Hydnum* and allies) 1.7	Surface of spines	Fleshy	241

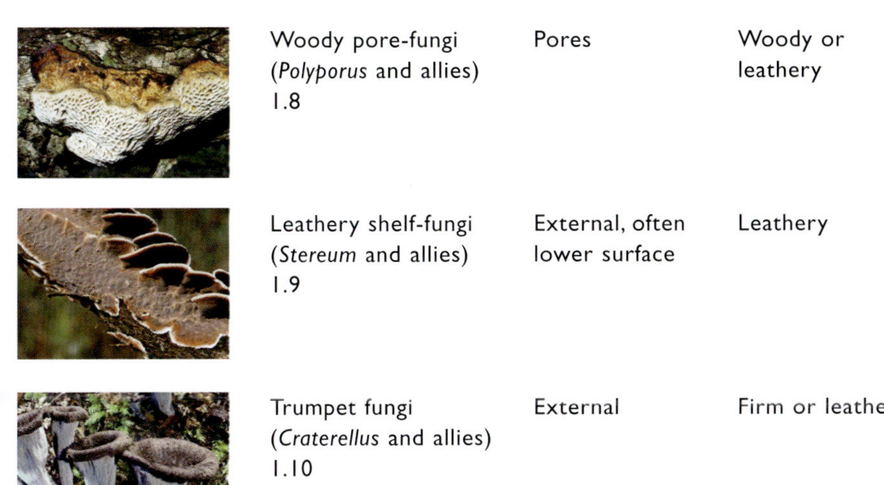

Group	Fertile surface	Texture	Page
Woody pore-fungi (*Polyporus* and allies) 1.8	Pores	Woody or leathery	245
Leathery shelf-fungi (*Stereum* and allies) 1.9	External, often lower surface	Leathery	276

Group	Fertile surface	Texture	Page
Trumpet fungi (*Craterellus* and allies) 1.10	External	Firm or leathery	287
Jelly fungi 1.11	External	Gelatinous	289

Section 2

ASCOMYCOTA

Group	Fertile surface	Texture	Page
Cup fungi and relatives 2.0	Within asci lining cups or discs or within embedded flasks	Fleshy to firm	297

Section 3

MYXOMYCOTA

Group	Fertile surface	Texture	Page
Slime moulds 3.0			343

Section 1
BASIDIOMYCOTA

This large section includes the fungi commonly referred to as mushrooms and toadstools. Most species are fleshy, as is the common field mushroom, and tend to decay quickly. However, some have a thin, relatively tough texture that persists for longer periods. The stems can be central or to the side. The gills are covered with fertile spore-bearing tissue that assumes the colour of the mature spores.

Agaricus is the mushroom genus to which the common field mushroom and its relatives belong. The white *Agaricus bisporus* is the most commonly available cultivated mushroom, but several other *Agaricus* species are cultivated commercially. There are many native species in Australia and several are edible. Some species are mildly poisonous, including the Yellow Stainer mushroom. Inedible species, including the Yellow Stainer, often give off a warning phenolic or disinfectant odour when heated or cooked. Most species have pink to greyish-pink gills when young, becoming dark brown with mature spores.

1.1 FUNGI WITH SIMPLE GILLS: Agaricus and relatives

1 *Agaricus augustus*

Caps to 200 mm across, sometimes larger, deeply convex to ovoid at first, becoming broadly convex to flattened with age. Immature specimens at first reddish brown, the surface skin breaking into a pattern of flat scales as the caps expand. The gills are pale or cream-coloured when young, ageing to dark brown. Gregarious, or forming caespitose clusters on the ground in forest or woodland. This species is edible in spite of a faint unpleasant odour when raw. **Spores** c. 8.5 x 9.5 μm, ellipsoidal, smooth. Spore print dark brown

2 *Agaricus austrovinaceus*

Caps to 100 mm across, nearly hemispherical at first, becoming flattened, undulate, the edges often upturned with age. Caps covered with appressed vinaceous to vinaceous-brown fibrils. Solitary to gregarious in eucalypt forest and occasionally pine forest. **Spores** c. 6 x 4 μm, sub-ellipsoidal, smooth. Spore print dark brown.

3 *Agaricus xanthodermus* Yellow Stainer

Caps to 100 mm across, more easily recognized before the caps expand. The young caps are usually white with a squarish appearance, with relatively long stems. All parts rapidly turn yellow if bruised or scratched. Young gills pale, soon becoming bright pink, finally dark brown. Mature caps pinkish white, flattened, and the yellow staining reaction is lost. If cooked, a phenolic or disinfectant-like odour is produced. Common in parks, lawns and forest edges. POISONOUS to most people. Two colour forms illustrated, (**3a, 3b**). **Spores** c. 7 x 5 μm, ellipsoidal, thick-walled, smooth. Spore print dark brown.

2

3a

3b

4 *Agrocybe parasitica*

Caps to c. 120 mm across. Convex at first, becoming broadly convex. A large membranous annulus is attached high up on the stem and is white at first, soon becoming chocolate-brown with the spores deposited from the gills. This species is caespitose on wood, usually on dead standing tree trunks in wet forest and rainforest, where it forms dense clumps. **Spores** c. 10 x 6 µm, ellipsoidal, smooth with thick walls and germ-pore. Spore print dark brown.

5 *Agrocybe praecox*

Caps to c. 70 mm across, convex to broadly conical with rounded apex, becoming flattened with age. The annulus, barely discernable in the photograph, is usually much larger. This is probably an introduced species; it is sometimes abundant on wood-chip mulch in parks and gardens. It is a common Northern Hemisphere species. **Spores** c. 9 x 6 µm, ellipsoidal, smooth. Spore print dark brown.

6 *Amanita ananiceps*

Caps to 70 mm across, covered with a layer of grey powdery meal and soft deciduous warts. The young convex caps become flat with age, losing the mealy layer to become smooth and shiny. The mealy, ragged fragments hanging from the cap margin are the remains of the annulus. In many characters this species is similar to *Amanita farinacea* (**8**), but is less variable in size and is grey instead of white. Usually found in eucalypt forest. **Spores** c. 8 x 6 µm, ellipsoidal to ovoid, smooth. Spore print white.

7 *Amanita cinereoannulosa*

Caps to c. 60 mm across, the stems to 100 mm long. A slender species with distinctive torn patches of tissue on the stem above and below the felty, grey annulus. Found in mixed coastal *Eucalyptus/Leptospermum* forest. Uncommon. **Spores** c. 11 x 8 µm, ellipsoidal, smooth. Spore print white.

8 *Amanita farinacea*

Caps to 100 mm across but usually c. 60 mm, broadly convex, becoming flatter with age. The whole mushroom is at first covered with a thick layer of powdery meal that quickly erodes. The pendulous rim of broken tissue on the margin of the cap is the remains of the annulus. The soft, conical warts on the young caps fall away with age leaving the surface smooth, white and satin-like. Common in eucalypt forest and woodlands. **Spores** c. 9 x 6.5 µm, broadly ellipsoidal to ovoid, smooth. Spore print white.

9 *Amanita grisella* var. *luteolovelata*

Caps to 80 mm across, at first almost hemispherical and covered with a rather thick, yellow, mealy universal veil which gradually erodes with age, leaving the cap patchy, finally smooth and shiny. Frequent in eucalypt forest and woodland. **Spores** c. 9 x 7 µm, ovoid, smooth. Spore print white.

10 *Amanita muscaria* Fly Agaric

Caps to 200 mm across. The striking scarlet caps are now a familiar sight under a variety of introduced trees (pines, birches and beeches, etc.). This exotic fungus, introduced from the Northern Hemisphere, has now become associated with native trees, mainly *Nothofagus* in Tasmania and Victoria, invading undisturbed rainforest, where it appears to be displacing indigenous mycorrhizal fungi. Distorted forms are often found. (**10a, 10b**). POISONOUS. **Spores** c. 10 x 7 μm, ellipsoidal, smooth. Spore print white.

10

10a

10b

11 *Amanita ochrophylla*

Caps to 200 mm across, convex, becoming broadly convex to flattened with age and with adhering patches of the universal veil. The gills are dull creamy yellow; the stem has a bulbous base with the annulus attached just below the gills. The fungus has an unpleasant, stale, ant-like odour. Frequent in eucalypt forest, particularly on roadsides. **Spores** c. 10 x 9 µm, subglobose, smooth. Spore print white.

12 *Amanita ochrophylloides*

Caps to 200 mm across, rounded at first, becoming flattened. Cap surface covered with large pointed warts that darken with age. Gills pale golden-yellow; annulus large and stem base bulbous. The odour is unpleasant, like the related *A. ochrophylla* (**11**). Frequent in open forest and woodland. **Spores** c. 8 x 7.5 µm, subglobose, smooth. Spore print white.

14

15

13 *Amanita phalloides* Death Cap

Caps to 120 mm across, at first almost globose and enveloped in a membranous universal veil. Expanding caps rupture the membrane and become broadly convex to flat. Colour variable with yellow-green, pale olive-green and yellow-brown forms most common, a white variety is less common. Introduced to Australia, where usually found under mature oaks. DEADLY POISONOUS. **Spores** c. 9 x 6.5 µm, subglobose, smooth. Spore print white.

14 *Amanita punctata*

Caps to 100 mm across, sometimes larger, almost subglobose at first, smooth and slightly viscid with pale, striate margin. Expanded caps are broadly convex, becoming flattened with a deeply sulcate margin. Distinctive characters include the large sack-like volva at the base of the stem, the lack of universal veil remains on the cap and the lack of an annulus. Found in moist forest, but uncommon. **Spores** c. 14 x 12 µm, globose to subglobose, smooth. Spore print white.

15 *Amanita umbrinella*

Caps to c. 70 mm across, grey or grey-brown covered with greyish, thin, felty patches that erode, leaving the cap with a satin-like lustre. The annulus is prominent and strongly striated. Fruit bodies appear in forested country after heavy rain at any time of the year, although they are more common in autumn and winter. **Spores** c. 7 x 6 µm, subglobose, smooth. Spore print white.

16 *Amanita xanthocephala*

Caps to 60 mm across. A small to medium-sized *Amanita* with a yellow to orange or red cap. The yellowish patches of veil remnants on the caps are readily washed away by rain. The usual *Amanita* annulus is fugacious in this species. A common species of forest, woodland and heathland. **Spores** c. 9 x 7 µm, ellipsoidal to subglobose, smooth. Spore print white.

17 *Anthracophyllum archeri*

Caps to 40 mm across, shallow-convex with reddish brown upper surface. The under surface is a bright brownish red, with few widely separated gills and a short lateral stem. A very distinctive species found on dead logs, branches and twigs in moist forest habitats. Fairly common. **Spores** c. 8 x 4.5 µm, ellipsoidal, smooth. Spore print white.

18 *Armillaria fumosa*

Caps to 100 mm across, convex, becoming flattened and finally depressed to funnel-shaped. The grey squamules on the cap and stem are fairly distinctive for the species. The specimens shown in the photograph were growing on the roots of *Eucalyptus ovata* in a swampy depression. **Spores** c. 7.5 x 5.5 µm, ellipsoidal, smooth. Spore print white.

19 *Armillaria hinnulea*

Caps to 90 mm across, broadly convex, finally flattened or somewhat centrally depressed and ornamented with fine brown warts. Gills rather close, creamy pink to greyish pink when mature. It is usually gregarious but sometimes forms small clusters on decaying logs and branches and sometimes on living tree trunks. This fungus is a tree parasite, usually found in wet mixed forest and rainforest. **Spores** c. 8 x 5 µm, broadly ellipsoidal, almost smooth. Spore print pale cream.

18

19

20 *Armillaria luteobubalina* Honey Fungus

→ Caps to 100 mm across, sometimes more, convex, becoming broadly convex to flattened and covered with a pattern of small wart-like scales. The stems are dark and firm. Size of the fruit bodies varies according to the size and density of the colony. It is an aggressive plant parasite, spreading from infected plants by underground string-like rhizomorphs (**20b**) to invade adjacent plants through their roots. It is a common forest fungus, but may invade gardens and orchards, where it can do much damage. **Spores** c. 9 x 6 µm, ellipsoidal, smooth. Spore print white.

21 *Armillaria novaezelandiae*

Caps to 100 mm across, plano-convex when mature, viscid. The stem is also viscid below the ring, becoming darker downwards. Found in clumps on dead wood and living trees in wet forests. A parasite of trees and other plants. **Spores** c. 8 x 5 µm, ellipsoidal, smooth. Spore print white.

22 *Armillaria pallidula*

Caps to c. 60 mm across, striate, minutely scurfy and slightly viscid at first. The gills are cream at first, ageing to pale pink. A tropical species usually found in clumps on dead tree stumps and sometimes arising from the underground roots of dead or living trees. **Spores** c. 8 x 5.5 µm, fairly variable, obovoid, smooth. Spore print white.

23 *Asterophora mirabilis* [= *Nyctalis mirabilis*]

Caps to 30 mm across. This species is only found growing on other fungi in the genus *Russula*. It grows in clusters and tends to prevent the host from rotting. Internal structure includes large, inflated cells (chlamydospores). On *Russula* spp. in wet forest and rainforest. Uncommon. **Spores** c. 6 x 4 µm ellipsoidal, smooth. Spore print white.

22

23

24 *Bolbitius muscicola*

Caps to 50 mm across, viscid, with a distinctive pattern of dark depressions. Gills and spores orange. It is found on the ground in cool temperate rainforest. It is related to the genus *Pluteus*, the species of which have pink spores. **Spores** not measured. Spore print white.

25 *Bolbitius titubans* [*Bolbitius vitellinus*]

Caps to 50 mm or more across, roundly conical at first, becoming broadly conical, then flattened. A fragile, watery species found on manure, decaying grass-straw and sometimes on lawns. Cosmopolitan. **Spores** c. 12 x 8 µm, ellipsoidal with germ pore. Spore print brown.

24

25

26 *Campanella gigantospora*

Caps to 30 mm, broad with very short lateral stem. The illustrated specimen was found on the bark of a standing tree, but the species may occur in other habitats. Uncommon. **Spores** c. 20 x 18 µm, subglobose. Spore print white.

27 *Campanella junghuhnii*

Caps to 15 mm across, growing in dense overlapping clusters on small dead branches and twigs in moist habitats. **Spores** c. 8 x 4.5 µm, smooth. Spore print white.

28b

28

29

31

28 *Campanella olivaceonigra* [= ***Tetrapyrgos olivaceonigra***]

Caps to 20 mm across. The stem and part of the upper cap surface are often blackish green. Found in loose colonies on dead twigs, small branches and dead sedges (**28b** shows under side). **Spores** 9 x 5 µm, rounded triangular in outline with central protuberance.

29 *Cantharellus concinnus* [= ***Cantharellus cibarius* var. *australiensis***]

Caps to c. 40 mm or more across and may be bright orange or pinkish orange with the gills running down most of the stem length. This fungus has a faint but distinctive odour and taste of apricot. Found in heathland and various forest types, but not common. **Spores** c. 8 x 6 µm, ovoid, smooth. Spore print white.

30 ***Cheimonophyllum candidissimum***

Caps to 10 mm across, pure white and very fragile. It could be confused with immature caps of *Crepidotus variabilis* (**66**), which is larger at maturity with pale brown gills. Found on decaying wood in wet forest. **Spores** c. 6 x 5.5 µm, subglobose, smooth. Spore print white.

31 ***Clitocybe clitocyboides***

Caps to 70 mm across, funnel-shaped, smooth with waxy texture and appearance. Colour ranges from cream to warm brownish cream. Gills decurrent. The presence of inflated hyphal cells in the cap tissue is a distinctive microscopic character. Found in moist to wet eucalypt forest and rainforest. **Spores** c. 6.5 x 3.5 µm, ellipsoidal, smooth. Spore print white.

32 *Clitocybe phyllophila*

Caps to 50 mm across, shallowly convex with central depression, the margin inrolled in young caps. Gills deep cream in mature specimens. Gregarious in pine forests among decaying pine needles. Coloured spores are unusual in the genus. **Spores** c. 5 x 3.5 µm, ellipsoidal, smooth. Spore print pale orange.

33 *Clitocybe semiocculta*

Caps to 70 mm across, white, becoming cream. Young caps are shallowly funnel-shaped with inrolled margins, becoming flatter with age. The fungus grows on fallen bark and wood debris. The name *semiocculta* refers to the half-hidden habit of the caps. **Spores** c. 7.5 x 5.5 µm, ellipsoidal, smooth. Spore print white.

34 *Collybia butyracea* [= *Rhodocollybia butyracea*]

Caps to 80 mm across, convex at first, becoming flattened. Surface texture greasy to touch, hygrophanous. Gills white, crowded, almost free from the stem. The colour of the young button stage is very variable, ranging from pale brown to chestnut brown. Common and widespread in forest and woodland. **Spores** c. 8 x 3.5 µm, ellipsoidal to almond-shaped, smooth. Spore print white to cream-coloured.

35 *Collybia dryophila* [= ***Gymnopus dryophilus***]

Caps to 30 mm across, conico-convex at first becoming flattened with central umbo, smooth, greasy to touch when moist, translucent-striate. Gregarious to caespitose on decaying eucalypt logs and stumps. **Spores** c. 6 x 3 µm, ellipsoidal to tear-drop shaped, smooth. Spore print white.

36 *Collybia eucalyptorum*

Caps to 25 mm across, convex to broadly convex, often with upturned margin, the surface dry, matt. Gills cream-coloured. Stem pale brown to reddish brown. Gregarious and common on fibrous bark of mature eucalypts in moist to wet forest. **Spores** c. 7 x 4.5 µm, ellipsoidal, smooth. Spore print white.

37 *Conchomyces bursiformis*

Caps to c. 100 mm across, semicircular to kidney-shaped and attached at the base without a stem. Young caps hoof-shaped, grey, with a pattern of white fibrils. It differs from *Crepidotus* in having white gills and spores. Gregarious or clustered on dead hardwood in moist to wet forest. **Spores** c. 7 µm, globose, covered with short spines. Spore print white.

38 *Conocybe filaris*

Caps 10–15 mm across, broadly conical, translucent-striate and granulose at apex. Stem slender with membranous annulus. A widespread species found on decaying wood in various forest habitats. POISONOUS. **Spores** variable c. 10 x 5 µm, ellipsoidal, smooth. Spore print yellow-brown.

39 ***Conocybe tenera***

Caps to 20 mm across, conical to bell-shaped, translucent-striate and slightly viscid when moist. Gill colour similar to cap. Stems tall and slender. Usually found on herbivore dung, sometimes on heavily manured soil. **Spores** c. 18 x 10 µm, ellipsoidal, smooth, with germ-pore. Spore print dark brown.

40 *Coprinellus disseminatus* [= *Coprinus disseminatus*]

Caps to c. 12 mm across, often appearing suddenly in large troops near old tree stumps and wooden stepping blocks, or from underground decaying roots. The emerging caps are creamy yellow, becoming grey when expanded and finally darken as the caps auto-digest. **Spores** c. 10 x 6 μm, almond-shaped, smooth, with germ pore. Spore print black.

41 *Coprinellus truncorum* [= *Coprinus truncorum*]

Caps to 35 mm across, conical, becoming somewhat irregular and mutually compressed as the dense colonies expand. This species closely resembles *C. micaceus* and is separated mainly on microscopic characters. Both species have small white flecks on the cap surface. Dense colonies occur around old stumps and on rotting wood. Successive crops may appear at the same location when the moisture conditions are suitable. **Spores** c. 8 x 5 μm, ellipsoidal, smooth, with germ pore. Spore print black.

42 *Coprinellus* sp.

Caps to 10 mm across, shortly cylindrical with convex apex, pale creamy beige and densely covered with relatively large hair-like cells that distinguish it from similar *Coprinus* species. The pale grey colour of the gills is also distinctive. Found on dead wood in moist gullies. **Spores** not examined.

41

42

43 **Coprinopsis nivea** [= *Coprinus niveus*]

Caps to 35 mm across, ovoid and nearly white when young, becoming bell-shaped then flattened. Grows on herbivore dung in moist places, singly or in small groups. Young specimens are covered with white powdery meal. The caps rapidly auto-digest to a black fluid. **Spores** c. 18 x 10 µm, ellipsoidal, smooth with germ pore. Spore print black.

44 *Coprinus atramentarius* [= ***Coprinopsis atramentarius***] Inky Cap

Caps to c. 80 mm across, deeply conical, becoming flared or bell-shaped as they expand. This species sometimes forms dense caespitose clumps that are strong enough to push up through asphalt and other prepared surfaces, causing considerable damage. Young caps darken and finally auto-digest to an inky fluid. Although this species is edible, it produces severe allergic reactions if consumed with alcohol. Common, often along road edges. **Spores** c. 8.5 x 6 μm, ellipsoidal, smooth, with germ-pore. Spore print black.

45 *Coprinus comatus* Shaggy Ink Cap or Lawyer's Wig

Caps to 70 mm across, at first ovoid, then long-ellipsoidal to variously bell-shaped. The caps soon turn black and liquefy, finally only a small black disc remains at the apex of the long white stem, which may be 300 mm or more high. An easily recognized fungus, frequently seen along road edges and on well-manured soils and rubbish tips rich in decaying organic material. Edible before the gills darken. Avoid consuming *Coprinus* with alcohol. **Spores** c. 12 x 8 μm, variable, ovoid, smooth, with germ pore. Spore print black.

46 *Cortinarius abnormis*

Caps to 70 mm across, convex at first, becoming plano-convex to plane, sometimes with small umbo. Widespread and fairly common in dry and moist eucalypt forest, often forming large open rings under eucalypts. The yellow-brown colour is fairly distinctive, but shaded gully forms tend to be darker. **Spores** c. 11 x 7 μm, variable, almond-shaped to rather lemon-shaped, warty. Spore print rust-brown.

47 *Cortinarius alboviolaceus*

Caps to 70 mm across, convex at first, becoming flattened with age. The pale, satiny, lilac caps contrast with pale cinnamon gills. A widespread species from eucalypt and mixed forest, usually gregarious, but occasionally caespitose. **Spores** c. 9.5 x 5 μm, ellipsoidal, minutely warty. Spore print rust-brown.

48 *Cortinarius archeri*

Caps to 100 mm across, convex at first, becoming broadly convex to flattened or slightly upturned. Cap, and stem below the membranous ring, are very glutinous when moist, and dry with a satiny sheen. The bright purple caps gradually fade to purplish brown and finally brown. Widespread, and fairly common in eucalypt forest. **Spores** c. 14 x 10 μm, almond-shaped, finely warty. Spore print rust-brown.

49 *Cortinarius areolatoimbricatus*

Caps to 150 mm across, deeply convex at first, becoming broadly convex and irregular at maturity. Cap surface is generally covered with overlapping, fibrillose scales. A robust, fleshy fungus usually found in dense caespitose clumps with caps distorted by mutual pressure. Common and widespread in eucalypt forest. **Spores** c. 10 x 5 μm, ellipsoidal, finely warty. Spore print rust-brown.

48

49

50

51

50 *Cortinarius australiensis*

Caps to 200 mm or more across, usually subglobose when young and covered by a tough membranous veil that breaks up into torn ragged fragments around the stem as the caps expand. Cap somewhat viscid when young, drying with a satin sheen. Fruit bodies often form massive caespitose clumps in eucalypt forest. **Spores** c. 12 x 6.5 μm, almond-shaped, warty. Spore print rust-brown.

51 *Cortinarius austroalbidus*

Caps to 70 mm across, convex at first becoming flattened with low central umbo, glutinous when wet, satiny when dry. Gills pale buff at first, darkening with age. Odour faintly curry-like. Widespread in eucalypt and mixed forest. **Spores** c. 10 x 6.5 μm ellipsoidal, warty. Spore print rust-brown.

52 *Cortinarius erythraeus*

Caps to 50 mm across, convex, viscid with inrolled margins, becoming irregular with maturity. The gills are at first pale olive, becoming brown with maturing spores. The lower part of the white stem is ornamented with red-brown fibrils. Frequent in eucalypt forest. **Spores** c. 10 x 7 μm, ellipsoidal, finely warty. Spore print rust-brown.

53 *Cortinarius* sp. aff. *largus*

Caps to 150 mm across, deeply convex at first, becoming broadly convex to flattened and irregular with age. The surface is smooth at first and slightly viscid, becoming minutely scaly with age. Gregarious in mixed eucalypt forest. **Spores** c. 10 x 6 μm, ellipsoidal to almond-shaped, warty. Spore print rust-brown.

52

53

55

54 *Cortinarius lavendulensis*

Caps to 90 mm across, broadly convex, sometimes undulate, glutinous when wet, pale lilac with light brown tints. Gills close, with lilac tinge. Stems usually stout, with a bulbous base, whitish with lilac tints. Gregarious in eucalypt forest. **Spores** c. 10.5 x 6 µm, almond-shaped, finely warty. Spore print rust-brown.

55 *Cortinarius rotundisporus*

Caps to 70 mm across, convex at first becoming flattened, viscid. The distinctive steely blue cap is often tinged with yellow or greenish yellow at the centre. The stem is relatively slender. It is widespread and fairly common in eucalypt forest and rainforest. **Spores** c. 8.5 x 6.5 µm, broadly ellipsoidal to subglobose, finely warty. Spore print rust-brown.

56 *Cortinarius sinapicolor*

Caps to 80 mm across, deeply convex at first, finally becoming irregularly flattened or somewhat upturned. The fruit bodies are easily recognized by the rich, bright yellow colour and the thickly glutinous caps and stems. A common species, often growing in discontinuous rings or arcs in various forest habitats. It has a mild, curry-like odour. **Spores** c. 8 x 5 µm, ellipsoidal, finely warty. Spore print rust-brown.

54

56

57 *Cortinarius sublargus*

Caps to 150 mm across, broadly convex to plano-convex, slightly viscid when moist, almost smooth. Gills at first cream, becoming darker with age. A distinctive character of this fungus is the short stem above ground with a deeply rooting underground portion descending to 200 mm or more into the ground. Frequent in eucalypt forest, more common after bush fires. An immature specimen showing cortina is depicted in **57b**. **Spores** c. 10 x 6 µm, ellipsoidal, warty. Spore print rust-brown.

58 *Cortinarius sp. aff. violaceus* [= ***Cortinarius austroviolaceus***]

Caps to 80 mm across, convex, becoming broadly convex to flattened, finely fibrillose, deep purple-violet. Gills fairly close, dark violet to almost black, becoming discoloured with spores. Stem, cortina and flesh also violet. Occasional in various eucalypt habitats. **Spores** c. 9 x 5 µm, ellipsoidal, finely warty. Spore print rust-brown.

57

57a

59b

59 *Cortinarius* sp.

Caps to 90 mm across, deeply convex at first, covered with a membranous veil, becoming broadly convex, with persistent veil remnants. The colour range varies between colonies, ranging from pale yellowish brown tones to forms flushed with pale lilac (**59b**) or greenish shades. Stem bulbous. Caespitose in eucalypt forest. Widespread. **Spores** c. 9 x 5 µm, almond-shaped to somewhat lemon-shaped warty. Spore print rust-brown.

60 *Cortinarius* sp.

Caps to 80 mm across, convex at first, becoming broadly convex and finally flattened to undulate-depressed or with broad central umbo. Cap viscid to subglutinous with distinctive concentric zonation. Stem greasy, not viscid. Gills pale brown, becoming rust-brown with spores. Found in dry *Eucalyptus*/ *Kunzea* forests east of Melbourne; otherwise distribution unknown. **Spores** c. 14 x 8 µm, almond-shaped, coarsely warty. Spore print rust-brown.

61 *Cortinarius* sp.

Caps to 80 mm across. At first almost hemispherical, expanding to deeply convex. The gills are very dark when just expanding and remain so to maturity. The stems are deeply rooting. This uncommon and distinctive *Cortinarius* is found in wet sclerophyll forests in southern Victoria, but its distribution elsewhere is not known. The clumps of dark brown caps flecked with blackish markings resemble colonies of the large puffball *Pisolithus arhizus* (**337**). **Spores** c. 9 x 5.5 µm, almond-shaped to somewhat lemon-shaped, warty. Spore print rust-brown.

62 *Cortinarius* sp.

Caps to 100 mm across, broadly convex to plano-convex, covered by a thin viscid pellicle that dries unevenly, leaving irregular pale flecks over the surface. The drab, cryptic colouring renders this fungus almost invisible amongst the leaf litter. Found in *Nothofagus* forests, where uncommon. Identity uncertain because of smooth spores. **Spores** c. 10.5 x 6.5 µm, broadly ellipsoidal, smooth. Spore print rust-brown.

61

62

63 →

63 *Cortinarius* sp.

← Caps to 80 mm across when expanded, at first almost hemispherical, becoming deeply convex then flattened, with margin tending to remain incurved. Illustrated specimens are not fully mature. The viscid to subglutinous surface dries to a satin-matt texture. The ring is membranous on a white stem. Genus uncertain because of smooth spores. Found in mixed *Eucalyptus*/*Nothofagus* forest. **Spores** c. 9.5 x 5.5 µm, broadly almond-shaped, smooth. Spore print rust-brown.

64 *Cortinarius* sp.

Caps to 100 mm across, at first broadly convex, expanding to irregularly undulate with a distinctive smell of burnt hair. Found in mixed *Eucalyptus*/*Leptospermum* forest. **Spores** c. 12 x 6.5 µm, almond-shaped, some constricted, warty. Spore print rust-brown.

64

65

66

65 *Crepidotus nephrodes*

Caps to 65 mm broad and 45 mm radius attached by a short lateral stem, mostly flat at maturity but sometimes distorted by crowding. Caps occur singly or sometimes in small clumps and are watery and brittle when mature. Common on decaying logs and branches. **Spores** c. 7.5 x 7 µm, globose, minutely spiny. Spore print brown.

66 *Crepidotus variabilis*

Caps to 15 mm across. Quite variable in size and form. Young colonies of almost stemless caps are white, but the gills gradually darken to brown as the spores mature. Found on small branches and twigs in moist, sheltered places. **Spores** c. 6 x 3.5 µm, ellipsoidal, finely spiny, Spore print pinkish brown.

67 *Cuphocybe* spp. [= **67a & b *Cortinarius* sp.**]

The genus is a relative of *Cortinarius* and is distinguished from it by the fibrillose-scaly cap surface, yellow-brown gills and stems with densely fibrillose scales. Several species of *Cuphocybe* occur in *Nothofagus* forests of south-eastern Australia. The ellipsoidal **spores** are usually warty. The two species illustrated (**67a, 67b**) are c. 70 mm diam. and are apparently undescribed.

68 *Cyptotrama asprata*

Caps to c. 50 mm across, covered with distinctive pointed fibrillose scales when young, which fall off as the cap expands. Found on decaying wood in tropical and subtropical forest. **Spores** c. 8 x 6 µm, subglobose, smooth. Spore print white.

67b

68

69

70

69 *Cystoderma amianthinum*

Caps to c. 35 mm across, broadly convex, umbonate and granulose at first, becoming yellower as the granules erode or wash away. Gills and stem above the ring white, the stem below the ring densely squamulose when fresh. Fruit bodies grow amongst moss in shady places and forest margins. The species is very similar to *Cystoderma musicolum* but larger. **Spores** c. 7 x 4 µm, ellipsoidal, smooth. Spore print white.

70 *Cystolepiota* sp. [= ***Cystolepiota aff.sistrata***]

Caps to 15 mm across, covered with microscopic spherical cells that give the surface a powdery appearance. The remains of the veil hang from the rim of young caps. Gills are white and the slender stems, c. 1.5 mm thick, are pale above and brownish below with scattered, scurfy fibrils. Mature fruit bodies can be mistaken for a *Mycena*. Widespread in forest leaf litter. **Spores** not examined. Spore print white.

71 *Cystolepiota* sp. [= ***Cystolepiota aff.adulterina***]

Caps to 15 mm across, densely covered with a thick layer of flocculent white powder, which is shed by the slightest vibration or breeze, forming a powdery deposit at the base of the slender stems. The cells forming the flocculent layer are relatively large and irregular in shape and frequently branched. The identity of this fungus is uncertain. It appears to be confined to decaying *Nothofagus cunninghamii* (myrtle beech) wood. **Spores** c. 10 x 5 µm, ellipsoidal, finely and densely punctate. Spore print white.

72 *Dermocybe austroveneta* [= ***Cortinarius austrovenetus***]

Caps to 80 mm across, at first convex, becoming broadly convex to flattened. Colour variable, ranging from yellow-green to dark blue-green that instantly turns red with alkali. Gills yellow becoming rust-brown. Gregarious to caespitose in eucalypt forest, where common. **Spores** c. 10 x 7 µm, ellipsoidal, warty. Spore print rust-brown.

73 *Dermocybe canaria* [= ***Cortinarius canarius***]

Caps to 80 mm across, convex, becoming broadly umbonate. Whole fungus bright yellow when fresh. Stems variable, usually with bulbous base. In eucalypt forest and rainforest. Uncommon. **Spores** c. 8 x 4.5 µm, ellipsoidal, finely warty. Spore print yellowish rust-brown.

74 *Dermocybe clelandii* [= ***Cortinarius clelandii***]

Caps to 70 mm across, convex at first, becoming flattened to undulate with depressed, dark brown centre. The pale mustard-yellow stems sometimes have a bulbous base. Widespread from wet eucalypt forest to rainforest. **Spores** c. 10 x 7 µm, ellipsoidal, coarsely warty. Spore print rust-brown.

75 *Cortinarius austrocinnabarinus*

Caps to 60 mm across, at first convex, becoming broadly convex and finally undulate. The brilliant cinnabar-red colour of the cap and stem bands are distinctive field characters. Uncommon, known from a few mountain forest sites in Victoria and Tasmania. **Spores** c. 7 x 5 µm, ovoid to ellipsoidal, finely warty. Spore print rust-brown.

76 *Dermocybe sanguinea* [= ***Cortinarius kulus***]

Caps to 40 mm across, at first broadly conical, becoming broadly convex. Finely radially fibrillose with silky sheen. Gills dark red becoming rusty coloured from spores. Stems blood red, slender. A common fungus forming gregarious colonies in eucalypt forest. **Spores** c. 9 x 7 µm, ellipsoidal, warty. Spore print rust-brown.

75

76

77 *Dermocybe splendida* [= **Cortinarius persplendidus**]

Caps to c. 50 mm across, at first convex, becoming broadly umbonate. Surface dry and radially fibrillose with dark brown centre. Gills brilliant orange-red, becoming brown with developing spores. Stems yellow with irregular reddish banding and slightly bulbous base. Widespread in *Eucalyptus/Leptospermum* forest. **Spores** c. 8 x 6 μm, ellipsoidal, warty. Spore print rust-brown.

78 *Dermocybe* sp. [= ***Cortinarius magentiannulata***]

Caps to 80 mm across, at first convex, becoming flattened with age. Gills pale brown. Found in *Eucalyptus/Leptospermum* forest. **Spores** c. 8 x 7.5 µm, subglobose to broadly almond-shaped, warty. Spore print rust-brown.

79 ***Dermoloma* sp.**

Caps to 25 mm across, convex to broadly convex, moist to slightly viscid. Gills very pale pinkish grey. Stem slightly tapering downwards, with a pattern of fine dark squamules just below the gills. Similar to the genus *Tricholoma*, but the cap surface consists of inflated cells instead of the long hyphae found in the latter. In dry eucalypt forest. Uncommon. **Spores** variable, c. 6.5 x 4 µm, ellipsoidal, smooth. Spore print white.

79

80 *Dermoloma* sp.

Caps to 35 mm across, nearly flat at first, becoming undulate, dry, grey with powdery bloom and pale margin. In dry eucalypt forest. Uncommon. **Spores** c. 8 x 5.5 µm, ovoid, smooth. Spore print white.

81 *Descolea recedens*

Caps to 40 mm across, convex, flattening with age. The soft fibrillose scales on the cap are the remains of a universal veil. Stems with a soft, membranous annulus. The genus is related to *Cuphocybe* and *Cortinarius*. Widespread, from heathland to rainforest. **Spores** c. 12 x 6.5 µm, oblique with pointed ends. Spore print rust-brown.

82 *Entoloma aromaticum*

Caps to 20 mm across, conical at first, then broadly conical with a darker, pointed umbo. Colour slightly variable, mostly brownish yellow but occasionally pinkish brown. Fruit bodies have a strong, distinctive, fruity odour. A widespread fungus of moist eucalypt forest and rainforest. **Spores** c. 10 x 7 μm, angular. Spore print pink.

83 *Entoloma moongum*

Caps to 45 mm across, broadly convex with central depression, finely radially fibrillose. The dark brownish black colour may appear bluish black, somewhat shiny and minutely rough. Gills at first pale purplish grey, becoming pale pink. Gregarious in eucalypt forest and woodland. **Spores** c. 11 x 7 μm, angular-ellipsoidal. Spore print pink.

82

83

84 *Entoloma rodwayi*

Caps to 30 mm across, convex to broadly conical, sometimes with flattened to depressed centre. On drying the yellow-green cap colour goes through a remarkable change to an intense blue-green. It occurs mainly in wet mixed forest and cool temperate rainforest. **Spores** c. 10 x 7 µm, angular, smooth. Spore print pink.

85 *Entoloma viridomarginatum*

Caps to 25 mm across, convex to broadly conical, with a small central depression. An important field character is the dark green edge to the gills. Cap colour is somewhat variable, from yellow-green to brilliant deep blue-green. Single to gregarious, in moss or grass on forest floor. Widespread. **Spores** c. 10 x 6 µm, angular. Spore print pink.

86 *Entoloma* sp.

Caps to 22 mm across, unevenly convex, translucent-striate and smooth. This distinctive species is characterized microscopically by a cap surface consisting of large globose cells. An uncommon species of open woodland. **Spores** c. 15 x 10 µm, angular. Spore print pink.

85

87

88c

88b

87 *Entoloma* sp.

Caps to 85 mm across, broadly conical at first, becoming plano-convex and somewhat undulate with low central umbo, radially fibrillose. Stems pale greyish blue. An unusually large *Entoloma*. Solitary to gregarious on ground in wet and dry eucalypt forest. Uncommon. **Spores** c. 9 x 7.2 µm, angular. Spore print pink.

88 *Favolaschia* spp.

The genus *Favolaschia* contains a small number of miniature fungi that belong to the Agarics, but have fertile surfaces with pores rather than gills. The white flared fruit bodies, up to 10 mm across, hang suspended from dead fern stems (**88a, 88b**). The small, cream-coloured species, c. 5 mm across, was growing on the under side of a dead eucalypt branch (**88c**). Found in wet forest and fern gullies. Both species are uncommon. **Spores** not measured. Spore print white.

89 *Filoboletus manipularis*

Caps to 20 mm across, with fine pores on the under side instead of gills. Flesh texture is thin and translucent. Gregarious to caespitose on decaying wood. Luminescent at night. A tropical to subtropical forest species. Related to Mycena. Common. **Spores** c. 7 x 5 μm, ellipsoidal, smooth. Spore print white.

88a

89

91

93

90 *Flammulina velutipes*

Caps to 100 mm across, usually growing in dense clusters. Stems of mature specimens are covered with a fine, dark brown, velvet-like layer. This species is edible and when cultivated is sold as enokitake. Cosmopolitan but, in Australia, usually found on trunks of *Acacia* species in wet forest. **Spores** c. 8 x 4 µm, ellipsoidal, smooth. Spore print white.

91 *Galerina hypnorum*

Caps to 10 mm across, bell-shaped, smooth, translucent-striate and very thin-skinned. This small, neat-looking species is always associated with mosses, sometimes on the ground or, more commonly, on decaying moss-covered logs in moist forest. Common. **Spores** c. 12 x 7 µm, almond-shaped, finely warty. Spore print yellow-brown.

92 *Galerina patagonica* [= ***Galerina marginata***]

Caps to 35 mm across, convex to broadly convex, slightly umbonate to centrally depressed, smooth, slightly viscid when moist, with translucent-striate edge. Gills with decurrent tooth. Stems firm, fibrillose-striate with a small membranous annulus. Gregarious to caespitose in pine forest on dead wood or on the ground amongst pine needles. **Spores** c. 8.5 x 5 µm, ellipsoidal to almond-shaped, finely warty. Spore print rust-brown.

93 *Galerina unicolor* [= ***Galerina patagonica***]

Caps to c. 40 mm across, convex, usually with an acute umbo. The whole fungus is of fairly uniform colour. It is common on dead wood, singly or in small colonies. A small ring is usually present on the stem. **Spores** c. 9 x 6.5 µm, almond-shaped, finely warty. Spore print yellow-brown.

92

94 *Gerronema postii* [= ***Loreleia postii***]

Caps to 25 mm across, convex to broadly convex with deep central depression. This fungus occurs with the thallose liverwort *Marchantia berteroana* and the moss *Funaria hygrometrica* after fires. It was first observed in several localities in burnt forest and heathland after extensive bush fires in Victoria in 1983. **Spores** c. 7.5 x 5 µm, ellipsoidal, smooth. Spore print white

95 ***Gymnopilus allantopus***

Caps to 50 mm across, at first conical becoming broadly conico-convex, flattened with age. Young caps have densely white-fibrillose incurved margins. Stems sometimes irregularly banded. Found on rotting wood in eucalypt and pine forests. **Spores** c. 8.5 x 6 µm, ellipsoidal, warty. Spore print mustard yellow.

96a

96b

96 *Gymnopilus dilepis*

Caps to 120 mm across, at first somewhat globose becoming broadly convex, then flattened. The reddish purple fibrils covering the caps and the yellow gills mimic the colour scheme of *Tricholomopsis rutilans* (**269**). The expanding caps become progressively paler owing to separation of the fibrils. This species appears to be confined to decaying pinewood, including sawdust and mulch, where it can form dense caespitose colonies. Immature fruiting bodies are shown in 96a, and expanded caps with surface scales are shown in 96b. **Spores** c. 7 x 4.5 µm, ellipsoidal, rough. Spore print yellow-brown.

97 *Gymnopilus junonius*

Caps to 400 mm across, finely fibrillose-scaly, subglobose at first, then deeply convex and finally irregularly flattened. Gills golden-yellow. Stems very bulbous when young. The fungus often forms spectacular, densely caespitose colonies on dead wood, particularly on and around stumps of harvested pines and hardwood trees. Sometimes a weak parasite of living trees. **Spores** c. 10 x 6 µm, ellipsoidal, warty. Spore print mustard yellow.

99

98

98 *Gymnopilus ferruginosus*

Caps to 70 mm across, convex to broadly convex, dry and covered with small, dark, fibrillose scales and with paler, mustard yellow margin. Gills cream at first, becoming dull golden-yellow. Common on decaying logs in eucalypt forest. **Spores** c. 8 x 5 µm, ellipsoidal, finely warty. Spore print yellowish brown.

99 *Hebeloma aminophilum* Ghoul Fungus

Caps to 120 mm across, ovoid at first, broadly convex to flattened with age, viscid, especially when young. The species grows on and near decaying animal remains including reptiles and birds. It is also found fruiting on the ground where animal carcases are buried, and occasionally at camp sites on decaying meat scraps and bones. **Spores** c. 9.5 x 5 µm, almond-shaped to lemon-shaped, warty. Spore print brownish pink.

100 *Hebeloma crustuliniforme* Poison Pie

Caps to 80 mm across, usually c. 60 mm, viscid or slimy when young. Flesh with radish-like odour. An introduced species, often common in pine plantations and among broad-leaved deciduous trees. POISONOUS. **Spores** c. 12 x 6.5 µm, almond-shaped, warty. Spore print dull brown.

101 *Hebeloma kammala*

Caps to 70 mm across, hemispherical at first, becoming broadly convex, viscid when young, sticky to touch when older. The pink gill colour is common to most *Hebeloma* species. Stems slightly expanded at the base. **Spores** c. 9 x 5 µm, ellipsoidal, finely warty. Spore print brownish pink.

102 *Hebeloma victoriense*

Caps to 150 mm across, deeply convex at first, becoming broadly convex and finally flattened, smooth, pale pinkish brown or cream, becoming darker with age. Gills bright pink. Stem with torn, ragged annulus usually coloured deep pink from deposited spores. Solitary to caespitose in eucalypt forest. **Spores** c. 12 x 7 µm, almond-shaped, warty. Spore print brownish pink.

103 *Hohenbuehelia atrocaerulea*

Caps to c. 25 mm across, mostly semicircular, attached laterally to dead logs and branches. The upper surface is pale creamy brown with scattered white fibrils. Large metuloidal cystidia are present on the gill face and edge (see endpapers for examples of microscopic characters). Most species of *Hohenbuehelia* possess organs that trap and digest nematodes. **Spores** c. 5 x 3.5 µm, ellipsoidal, smooth. Spore print white.

102

103

104 *Hohenbuehelia clelandii*

Caps to 80 mm across, up to 80 mm long, laterally attached to decaying logs, stumps and exposed dead roots. The upper surface is medium to dark brown and slightly viscid. Metuloidal cystidia are present on gills. **Spores** c. 8 x 5 µm, ellipsoidal, smooth. Spore print white.

105 *Hohenbuehelia petaloides*

Caps to c. 80 mm or more high, usually in clumps. The split, funnel-shaped fruit bodies are characteristic. As in other species in the genus, the fruiting body has a clear elastic, gelatinous layer beneath the cap surface. Metuloidal cystidia are present on the gills. Most frequent in pine plantations. **Spores** c. 8 x 4.5 µm, ellipsoidal, smooth. Spore print white.

106

107

106 *Hohenbuehelia* sp.

Caps to 60 mm across, flabelliform, viscid with a clear elastic gelatinous layer below the cap surface. Metuloidal cystidea are present on the gills. Found on the trunks and stumps of dead trees and on the ground, rising from buried dead roots. **Spores** c. 9 x 5.5 µm, ellipsoidal, smooth. Spore print white.

107 *Hygrocybe aurantiopallens*

Caps to 20 mm across, nearly hemispherical with strongly incurved margins when young, becoming broadly convex. Whole fruit body bright orange when young, apricot to apricot-yellow with age. Greasy to touch when moist. Occasional in wet or shaded eucalypt forest and rainforest. **Spores** c. 5 x 4 µm, usually teardrop-shaped, smooth. Spore print white.

108 *Humidicutis arcohastata*

Caps 20–40 mm across, conical to broadly conical, displaying a remarkable colour transition throughout their development. Emerging caps are deep purple-green, with margins becoming olive-green, changing to greenish yellow and finally orange-red. Gregarious to slightly caespitose in eucalypt forest. Uncommon. **Spores** c. 8.5 x 5.5 µm, ellipsoidal, smooth. Spore print white.

109 *Hygrocybe astatogala* [= ***Bertrandia astatogala***]

Caps to 50 mm across, conical, becoming broadly conical. Variable in colour, with tones of orange, yellow or red, with overlying black radial fibrils, the cap becoming wholly black with age. The stems, also with overlying black fibrils, exude a clear fluid when cut. Found in various forest types in moist habitats, but most common in rainforest. **Spores** c. 9 x 7 µm, ovoid, smooth. Spore print white.

110 *Hygrocybe austropratensis*

Caps to 50 mm across. An uncommon, rather robust *Hygrocybe* with distinctive thickened stem. Caps are convex when immature, becoming broadly convex and sometimes centrally depressed when fully expanded. Colour is somewhat variable. Two forms are illustrated (**110a, 110b**). Found in wet or dry eucalypt and mixed forest, frequently amongst moss. **Spores** c. 6.5 x 5 µm, ellipsoidal, smooth. Spore print white.

109

110a

110b

111 *Hygrocybe cheelii* [= ***Cantharellus lilacinus***]

Caps to 70 mm across but usually less, convex at first, becoming deeply funnel-shaped and frequently distorted when growing in clusters. The fold-like gills are strongly decurrent; the base of the stem is yellow. Occurs singly or in scattered colonies or occasionally in small clumps in wet or dry forest, often amongst moss. **Spores** c. 8.5 x 5.5 µm, ellipsoidal, smooth. Spore print white.

112 *Hygrocybe chromolimonea*

Caps to 20 mm across, with slightly crenulate margins. The whole fungus is viscid or slimy. Found in rainforest and in moist areas of eucalypt forest. Gregarious on decaying wood and leaf litter. **Spores** c. 8 x 5 µm, ellipsoidal, smooth. Spore print white.

111

112

113 *Hygrocybe coccinea*

Caps to 30 mm across. Gills pink. Found in moist to swampy habitats. **Spores** c. 8 x 5 μm, ellipsoidal, smooth, white. Spore print white.

114 *Hygrocybe conica*

Caps to 40 mm across, narrowly to broadly conical, slightly viscid when moist. Colour ranges through yellow, yellow-green, orange, red and finally black, with older caps radially splitting. Gregarious in moist, mossy areas. **Spores** c. 12 x 8 μm, ellipsoidal, smooth. Spore print white.

115 *Hygrocybe firma*

Caps 30 mm across, convex to broadly convex, dry. Gills pink, decurrent. A conspicuous species because of its brilliant colour. It has two different spore types. Found in eucalypt forest and rainforest in moss beds. **Spores** Larger spores c. 14 x 9 µm, smaller spores c. 8.5 x 5.5 µm, unevenly ellipsoidal, some constricted, smooth. Spore print white.

117

116 *Hygrocybe fuhreri*

Caps to 20 mm across, usually smaller. So far only known from the type locality in Victoria ,where it grows in small colonies in moss in *Kunzea/Eucalyptus* woodland. A distinctive but uncommon fungus. **Spores** c. 9 x 5 µm, ellipsoidal with slight constriction, smooth. Spore print white.

117 *Hygrocybe graminicolor* [= ***Gliophorus graminicolor***]

Caps to 25 mm across, with stem length about three times the cap diameter. Fruit bodies very glutinous and translucent. Colour changes to pink with age or drying. Found in wet forest and rainforest. **Spores** c. 6.5 x 4 µm, oblong, smooth. Spore print white.

118

119

120

118 *Hygrocybe leucogloea*

Caps to 40 mm across, flattened to irregularly saucer-shaped and almost white with pale brown centre. Cap surface viscid (note adhering soil particles.) Found mainly in eucalypt forest. **Spores** c. 6.5 x 4.5 µm, ellipsoidal, smooth. Spore print white.

121

119 *Humidicutis lewellinae* [= ***Porpolomopsis lewellinae***]

Caps to 60 mm across, conical, becoming flattened, usually with a low umbo. Caps tend to split radially through the gill axis when expanded. Usually found in coastal heath and woodland, but occasionally occurs inland in wet forest, where the fruiting body colour is more intense. Widespread. **Spores** c. 9 x 6 µm, variable, ovoid to ellipsoidal, smooth. Spore print white.

120 *Hygrocybe lilaceolamellata*

Caps to 20 mm across, with central depression in mature specimens. The specific name refers to the lilac gill colour. The spore deposit is pale violet instead of the usual white for the genus. Found in wet forest and rainforest. **Spores** c. 8.5 x 4.5 µm, oblong, constricted, smooth. Spore print pale violet.

121 *Humidicutis mavis*

Caps to 50 mm, at first conical, becoming plane with a raised central umbo at maturity. This species resembles a white form of *H. lewellinae*, and exhibits the same type of gill-splitting as the cap expands. Usually found in rainforest. **Spores** c. 7.5 x 5 µm, broadly ellipsoidal, smooth. Spore print white.

122

122 *Hygrocybe miniata*

Caps to 30 mm across, convex, becoming plane to undulate, moist, smooth. Gill colour similar to cap or with some yellow. Stem colour similar to cap, or with yellow tones at the base. Gregarious, sometimes caespitose. Usually on the ground in moist forest and rainforest, also in wet heathland. **Spores** c. 8.5 x 5 µm, broadly ellipsoidal, smooth. Spore print white.

123 *Hygrocybe persistens var. persistens* [= *Hygrocybe acutoconica*]

Caps to 45 mm across, conical, becoming broadly conical to flattened, with umbo. At first moist, becoming dry and radially fibrillose, radially splitting at maturity. The long stems are often buried for much of their length. Gills and stems yellow. Widespread and common on calcareous coastal sand dunes. **Spores** c. 12 x 8 µm, cylindrical, slightly constricted, smooth. Spore print white.

124 *Hygrocybe pseudograminicolor*

Caps to 15 mm or more across, with broad central depression. The whole fungus is viscid and the flesh is very fragile. The unusual colour makes this fungus unlikely to be confused with any other species. It is frequent along the east coast of Australia, especially in rainforest. **Spores** c. 10 x 6 µm, ellipsoidal, smooth. Spore print white.

123

124

125

126

128

125 *Hygrocybe psittacina* var. *perplexa* [= ***Gliophorus perplexus***]

Caps to 30 mm across, with both cap and stem glutinous. An uncommon and distinctive species found in moist forest amongst moss. **Spores** c. 7 x 4.5 µm, ellipsoidal, smooth. Spore print white.

126 ***Hygrocybe rodwayi***

Caps to 35 mm across, all white with decurrent gills. It may be confused with *H. virginea*, which has larger spores. A common species in eastern Australia. **Spores** c. 5.5 x 4.5 µm, subglobose, smooth. Spore print white.

127 ***Hygrocybe saltorivula***

Caps grow to 35 mm across, broadly conical at first, becoming flattened at maturity, with central umbo. The colour is not uncommon in the genus, but the microscopic characters are distinctive. Gregarious in moss in shaded *Eucalyptus/Kunzea* woodland. Rare. **Spores** c. 8.5 x 4.5 µm, ellipsoidal, constricted, smooth. Spore print white.

128 ***Hygrocybe stevensoniae***

Caps to 30 mm across, stem length c. 1.5 times the cap diameter. Similar to *H. graminicolor*, with glutinous cap and stem also turning pink on drying, but the stems are not as slender. Found in wet forest and rainforest habitats. Uncommon. **Spores** c. 9 x 5 µm, ellipsoidal, smooth. Spore print white.

129 *129a Hygrocybe virginea* [= ***Cuphophyllus virgineus***]
129b Cuphophyllus virgineus [= ***Cuphophyllus virgineus* var. *fuscesens***]

Caps to 30 mm across. Two varieties are recognized, both found occasionally in wet forest and rainforest.

129a, var. *virginea* is white with **spores** measuring c. 9.5 x 6.5 μm.

129b, var. *fuscescens* has yellow-brown tones and gills that are more decurrent, giving the cap a funnel shape, with **spores** c. 9.5 x 4.5 μm, ellipsoidal, smooth (narrower than in var. *virginea*). Both have white spore prints.

---

1 See A.M.Young in *Muelleria* 14:61-62 (2000). Two small collections from Warrandyte State Park are described but not validly named due to 'insufficient material for ...deposition of a suitable type'.

2 *Ibid.*

130 *Hygrocybe xanthopoda*

Caps to 50 mm across, conical at first, becoming flattened and undulate at maturity, often with a low umbo. The gills are thick and distant, pale at first, becoming yellow. The yellow to orange-yellow stems are relatively thick for the genus. Gregarious in wet eucalypt gullies or rainforest. **Spores** c. 9 x 6 µm, ellipsoidal, smooth. Spore print white.

131 *Hygrocybe* sp. JCR1 (Tony Young[1])

Caps to 20 mm across, viscid to slightly glutinous. Stems tapering downwards. An uncommon species found amongst moss under *Kunzea* thickets in dry eucalypt woodland. **Spores** c. 10 x 5 µm, ellipsoidal. Spore print white.

130

131

132 *Hygrocybe* sp. JCR 2 (Tony Young[2]) – see footnote page 92

Caps to 20 mm across, remaining more or less conical, with a moist, greasy texture. The gills are orange-pink. Found amongst moss in eucalypt woodland. Uncommon. **Spores** c. 10 x 5.5 µm, ellipsoidal, with slight constriction, smooth. Spore print white.

133 *Hygrocybe schistophila*

Caps to 25 mm across, convex with a central depression, becoming flattened-undulate, smooth, moist. Gills thick, distant, the longer ones decurrent. Gregarious to sub-caespitose amongst moist, stony rubble on the bank beside a mountain track in Tasmania. Rare, so far known only from the one locality. **Spores** c. 9.5 x 6 µm ellipsoidal, smooth. Spore print pale yellow.

134

135

134 *Hygrophorus involutus*

Caps to 30 mm across, broadly convex and glutinous. Gills cream, pale yellow or pale apricot. Distinctive characters include an inrolled cap margin and the presence of water droplets on the upper stem. Found in rainforest and moist eucalypt forest. **Spores** c. 6 x 3 µm, ellipsoidal, smooth. Spore print white.

135 *Stropharia aurantiaca* [= ***Leratiomyces ceres***]

Caps to 60 mm across, convex at first, then plane or with a low umbo. Colour varies from reddish brown to bright tomato-red. Gills at first grey, becoming purplish brown as spores mature. This species is most common on wood mulch in parks and gardens, where it often forms dense colonies. Cap colour seems to vary with substratum. **Spores** c. 12 x 7 µm, ellipsoidal with germ pore, smooth. Spore print purple-brown.

136 *Hypholoma brunneum*

Caps to 50 mm across, convex with a slightly raised umbo, viscid when moist and, when young, ornamented with scattered white fibrillose scales which are larger and more persistent around the margin. The brownish cream gills darken as the spores mature. Caespitose on dead wood in rainforest and wet eucalypt forest. **Spores** c. 7 x 4 µm, ellipsoidal, smooth. Spore print purple-brown.

137 *Hypholoma fasciculare*

Caps to 70 mm across with veil fragments adhering to the margins of young specimens. The gills are a distinctive greenish yellow when young, becoming darker with maturing spores. Found on tree stumps, dead logs and branches, usually in dense clusters. Common. POISONOUS. **Spores** c. 7.5 x 5 µm, ellipsoidal, smooth with germ pore. Spore print purple-brown.

136

138 *Hypholoma sublateritium* [= ***Hypholoma australe***] **Brick Caps**

Caps to 60 mm across, deeply convex at first, becoming broadly convex or flattened, often forming large spectacular clumps at the bases of living or dead trees, or on the ground rising from dead buried wood. The gills are yellow-green when young, becoming darker as the purple-black spores mature. The common name refers to the orange-brown or brick-red colour of the caps. **Spores** c. 7.5 x 4.5 µm, ellipsoidal with germ pore, smooth. Spore print purple-brown.

139 *Hypholoma* sp. [= ***Hypholoma fasciculare var. americana***]

Caps to 30 mm across, broadly convex with a prominent umbo. The gills are close and orange without any green tinge when young. Mature gills turn dark purple-brown with spores. Common on dead logs in wet forest and rainforest. **Spores** c. 7 x 4 µm, ellipsoidal with germ pore, smooth. Spore print purple-brown.

140 *Inocybe australiensis*

Caps to 20 mm across, convex with central umbo, surface covered with dark brown fibrillose scales. Stems pale brown to red-brown, squamulose with slightly bulbous base. Scattered on ground in heathland and forest. Widespread but not common. **Spores** c. 7.5 x 5 µm, broadly almond-shaped, smooth. Spore print dull brown.

139

140

141 *Inocybe austrofibrillosa*

Caps to 40 mm across, deeply convex at first, becoming broadly conical, some-times with an acute umbo. The cinnamon-buff surface is covered with white fibrils, giving the fungus a hoary appearance. The pale cinnamon, watery stem has scattered to dense white fibrils. Gregarious to caespitose in moist to wet eucalypt forest. **Spores** c. 9.5 x 5.5 µm, ellipsoidal, smooth. Spore print white.

142 *Inocybe eutheles*

Caps to 50 mm across, broadly bell-shaped, with a distinct umbo. Gills pale, becoming clay-brown when mature. Stem base slightly swollen. The fungus has a strong odour of chlorine bleach. Found mostly in pine plantations, often forming extensive colonies. **Spores** c. 9 x 4.5 µm, almond-shaped, smooth. Spore print dull brown.

143 *Inocybe* sp.

Caps to 70 mm across, broadly plano-convex and densely covered with concentric zones of loose fibrillose scales. The cap surface pulls away from the cap margin, exposing the smooth under surface. Cap margin remains inrolled until the cap is fully expanded. The stem is also covered with a dense fibrillose layer. Found on the ground, associated with *Nothofagus cunninghamii* (myrtle beech). **Spores** c. 10 x 5.5 µm, ellipsoidal, smooth. Spore print rust-brown.

144 *(a) Laccaria canaliculata* (left-hand group)

Caps to 40 mm across, convex becoming flattened or saucer-shaped with age, translucent-striate, hygrophanous. May be confused with related species. Widespread and common in forest and woodland. **Spores** c. 7 µm diam., globose, spiny. Spore print white.

144 *(b) Laccaria masoniae* (right-hand group)

Caps to 40 mm across. Stems slender, reddish brown, contrasting with the pale pink to pinkish grey caps. The unusually long stem is a distinctive character. This species is usually found in association with *Nothofagus cunninghamii* (myrtle beech). **Spores** 7–10 µm diam., covered with large, pointed spines. Spore print white.

144

145 →

145 *Laccaria lateritia*

← Caps to 30 mm across, broadly convex to flattened, sometimes centrally depressed. Margin often undulate and usually translucent-striate. Common in forest, woodland and heathland, also found in gardens. **Spores** c. 7–10 µm diam., globose or subglobose, covered with spines. Spore print white.

146 *Laccaria proxima*

Caps to 40 mm across, convex, becoming broadly convex, often centrally depressed. Margin sometimes undulate. Stems often twisted. Occurs singly or gregarious on the ground amongst conifers. **Spores** c. 8 x 7 µm, subglobose and covered with spines. Spore print white.

146

148

147 *Laccaria sp.*

Caps to 70 mm across, convex at first, becoming broadly convex to flattened, undulate or centrally depressed. Texture waxy, hygrophanous. Probably the largest Australian *Laccaria*. Common in cool temperate rainforest with *Nothofagus*. Gregarious to caespitose. **Spores** c. 9.5 μm diam., globose, covered with large pointed spines. Spore print white.

148 *Lactarius clarkeae*

Caps to 100 mm or more across, convex when young, becoming broadly depressed or funnel-shaped. A white milky fluid is produced when gill or cap tissue is cut or broken. Scattered in open forest and woodland. **Spores** c. 9 x 7 μm, globose with amyloid ridges. Spore print white.

149 *Lactarius deliciosus* Saffron Milk Cap

Caps to 160 mm across, occasionally larger. Stems have a distinctive pattern of darker orange blotches. When cut, the gills and flesh exude a carrot-coloured milky juice. When bruised or old, all surfaces become green. An exotic species associated with pines. Edible. **Spores** c. 9 x 7 µm, ellipsoidal, with amyloid warty ridges. Spore print pale yellow.

150 *Lactarius eucalypti*

Caps to 40 mm across, convex at first, becoming centrally depressed, finally funnel-shaped. Broken gills produce a white 'milk' with a mild to peppery taste. Common in eucalypt forests. **Spores** c. 8 x 6.5 µm, ellipsoidal, with raised amyloid reticulations. Spore print white.

151 *Lactarius piperatus* Peppery Milk Cap

Caps to 120 mm across, convex with central depression when young, becoming broadly funnel-shaped, producing an extremely hot, acrid, white milk when cut. Gills crowded, cream to pinkish cream. Found under introduced and native trees, and probably introduced. **Spores** c. 8 x 6 µm, ovoid with amyloid reticulations. Spore print white.

152 *Lactarius plumbeus* Ugly Milk Cap

Caps to 120 mm across, convex, becoming broadly convex with central depression. All parts react with alkali such as ammonia or potassium hydroxide solution to produce an instant purple-violet colour. The white latex produced on cutting slowly turns brown and tastes extremely hot and acrid. An introduced fungus associated with birch trees. **Spores** c. 9 x 7 µm, ellipsoidal with amyloid ornamentation. Spore print white.

151

152

155

153 *Lactarius wirrabara*

Caps to 80 mm across, dark chocolate brown with a fine, felty texture of microscopic, rigid, septate hairs. Stems also hairy. Gills distant, staining pink and producing a creamy white latex that slowly turns pale milky brown. Gregarious in eucalypt forest. Uncommon. **Spores** c. 8.5 x 8 µm, subglobose with amyloid reticulations. Spore print white.

154 *Lentinellus* aff. *omphalodes* [= ***Lentinellus tasmanicus***]

Caps to 40 mm across, shallow funnel-shaped with down-curved margin, smooth, undulate. Stem central, firm, sometimes woolly at apex. Gills very broad, deeply and unevenly serrated. Found on dead wood in moist forest. **Spores** c. 5 x 4 µm, broadly ellipsoidal, minutely warty.

155 *Lentinellus pulvinulus* [= ***Lentinellus castoreus***]

Caps to 50 mm across, usually clustered on dead wood. Gill edges are finely and unevenly toothed. Occasional on dead wood in wet forest. **Spores** c. 4.5 x 4 µm, subglobose, finely warty.

156 *Lentinellus* aff. *ursinus* [= ***Lentinellus pulvinulus***]

Caps to 40 mm wide and 30 mm radius, broadly convex to flattened, mostly finely tomentose but with smooth margin and coarse dark hairs around attachment. The cream gills are unevenly serrated. Found on dead wood and bark in moist forest. **Spores** c. 4 x 3.5 µm, subglobose, minutely spiny. Spore print white.

157 ***Lepiota aspera***

Caps to 100 mm across with prominent, sometimes pointed, scales that are usually crowded in the central region. The annulus is relatively large, but often tears and falls away. Uncommon, but widespread in wet forest. **Spores** c. 7.5 x 2.5 µm, cylindrical, smooth. Spore print pale yellow-brown.

158 *Lepiota haemorrhagica*

Caps to 60 mm across, with red-brown fibrillose scales. The annulus and stem are also red-brown. The expanding cap surface splits to reveal the paler flesh beneath. The gills are white and free from the stem and stain red with bruising or age. The whole fungus dries dark red-brown. A common forest species. **Spores** c. 7 x 5 µm, ellipsoidal, with germ pore, smooth. Spore print pinkish white.

159 *Lepiota* sp.

Caps to 50 mm across, broadly convex to flattened, with dark fibrillose umbo. Dark brown fibrils radiate from umbo over a pale background. The stem above the annulus is white and below has a fibrillose pattern. Found in open forest and woodland. **Spores** c. 11 x 5 µm, ellipsoidal to narrow-ellipsoidal, thick-walled, smooth. Spore print white.

158

159

161

160 *Lepiota* sp.

Caps to 25 mm across, broadly convex and nearly smooth, with low broad umbo, centre grey with a lilac tinge, becoming paler towards the margin. Stem translucent, with annulus approximately central. Found in open forest and woodland, usually in moist depressions. **Spores** c. 8 x 4 µm, ellipsoidal to cylindrical-ellipsoidal, thick-walled, smooth. Spore print white.

161 ? *Lepiota* sp.

Caps to 15 mm across, deeply convex. Whole fruit body ornamented with tiny amber, watery globules. This very delicate species prefers the corky remains of old rotten wood, but is occasionally found under decaying wood on the ground. Widespread but uncommon. **Spores** c. 7 x 4 µm, ellipsoidal, smooth. Spore print white.

162 *Lepista nuda*

Caps to 150 mm across, convex at first, becoming flattened to undulate. Fresh specimens are all lilac, the cap surface becoming brown with age. Early records from near pines and exotic trees suggest that it is an introduced species. It is becoming increasingly common and widespread in our native forests. It is a handsome species with a delicate scent, edible when cooked but reported as mildly toxic if eaten raw. **Spores** c. 8 x 5 µm, ellipsoidal, finely roughened. Spore print pale salmon-pink.

163

163 *Leucoagaricus naucinus* [= ***Leucoagaricus leucothites***]

Caps to 100 mm across, deeply convex at first, then broadly convex to flattened. The annulus is thick and membranous. This fungus resembles a field mushroom, but the cap and gills are white, turning cream with age. Gregarious in forest and heathland. **Spores** c. 9 x 6 µm, ellipsoidal, with small germ pore, smooth. Spore print white.

164 ***Leucoagaricus ooliekirrus***

Caps to 50 mm across, ovoid and mealy at first, becoming flattened and smooth with age. At first white, becoming pale buff at the centre when mature. Fruit body fragile. Common in moist forest. **Spores** c. 12 x 7 µm, ellipsoidal, with small germ pore. Spore print white.

165 ***Leucoagaricus rubrotinctus***

Caps to 60 mm across, covered with reddish brown radial fibrils. Gills free from the stem as in *Lepiota*. Common in forest and heathland. **Spores** variable, c. 8 x 5 µm, ellipsoidal, smooth. Spore print white.

164

165

166 *Leucocoprinus birnbaumii*

Caps to 60 mm across, ovoid at first, becoming bell-shaped then flattened and soon collapsing. This distinctive fungus is usually seen in flowerpots and planter boxes in nurseries, and outdoors in warm climates. **Spores** variable c. 10 x 7 µm, ellipsoidal, smooth, with small germ pore. Spore print white.

167 *Leucopaxillus amarus*

Caps to 140 mm across, convex with inrolled margin when young, becoming plano-convex and finally funnel-shaped. Surface scaly owing to the surface cracking as the caps expand. Darker surface patches are often present. Copious white mycelium occurs in the soil beneath the fruit bodies. Usually associated with introduced conifers. **Spores** c. 10 x 4 µm, subglobose with prominent, scattered amyloid warts. Spore print white.

166

167

168

168 *Leucopaxillus eucalyptorum*

Caps to 150 mm across, convex, becoming flattened to centrally depressed, sometimes with concentric zones of watermark-like patches. Gills crowded, with slightly decurrent stem attachment. This species is usually found under senescent or diseased trees with dense layers of shed bark at the base. The fruit bodies arise from a dense layer of white, basal mycelium. **Spores** c. 6 x 4.5 µm, ellipsoidal, ornamented with minute amyloid warts. Spore print white.

169 *Leucopaxillus lilacinus*

Caps to 100 mm across, occasionally larger, unevenly convex with inrolled margin, becoming flattened with incurved margin, almost smooth, lilac, becoming darker with age. Gills crowded, white, ageing to cream. Gregarious to caespitose, in eucalypt forest. Spores c. 6 x 5µm, ovoid, ornamented with strongly amyloid, uneven warts. **Spore** print cream-coloured.

170 *Limacella pitereka*

Caps to 70 mm across, convex, becoming plano-convex with low umbo, glutinous, usually with brownish centre. Gills creamy white. The lower stem is also glutinous. Found in eucalypt forest and woodland and occasionally in pine forest. The genus is related to Amanita, but both the universal veil and annulus are glutinous in Limacella. **Spores** c. 6 x 4.5 µm, subglobose, smooth. Spore print white.

171 *Limacella* sp.

An undescribed species similar to *L. pitereka*, but caps are light brown with a deeper brown centre. **Spores** not examined.

172 *Lyophyllum connatum* [= ***Leucocybe connata***]

Caps to 200 mm across, deeply convex at first, expanding with maturity. Forms dense, caespitose clumps with caps often deformed by mutual pressure. Usually found in warm to tropical regions in grass and forest edges. An immature colony is illustrated. **Spores** c. 7 x 4 µm, ellipsoidal, smooth. Spore print white.

173 ***Macrolepiota clelandii***

Caps to 120 mm across. Stems up to 200 mm high. Frequent in open forest and sometimes common along forest roadsides. **Spores** c. 18 x 12 µm, variable, ellipsoidal, smooth, with germ pore. Spore print creamy white.

174 ***Macrolepiota dolichaula***

Caps to 170 mm across, ivory-white with a pale brown umbo and pale buff scales. The stem is nearly twice as long as the cap diameter. Gills are cream. The annulus on the stem is movable. Usually occurs in warm climates in grassy places. **Spores** c. 14 x 10 µm, ellipsoidal, smooth, with germ pore. Spore print creamy white.

175 *Macrolepiota procera* [= ***Macrolepiota clelandii***]

Caps to 200 mm or more across, usually c. 150 mm. Stems to 200 mm high, often with darker bands. The large size and dark, almost shiny, central umbo distinguishes this species from its near relatives. Solitary to gregarious in pine forest and with other exotic trees. Presumably introduced. **Spores** c. 18 x 12 µm, ovoid, smooth. Spore print white.

172

174

173

175

176 *Marasmiellus affixus*

Caps to 15 mm across, with very short lateral stem. Fruit bodies form crowded, often extensive colonies on dead eucalypt bark and branches. It produces a dirty cream-yellow discoloration on the substratum and frequently cements small twigs and branches together with rhizomorphs. It has a strong, unpleasant odour. **Spores** c. 7 x 4 µm, ellipsoidal, smooth. Spore print white.

177 *Marasmiellus candidus*

Caps to c. 15 mm across, at first broadly funnel-shaped, white but sometimes becoming faintly pinkish with age. The species colonises dead wood, usually small standing trunks of various tree species. In moist to wet forest. **Spores** c. 14 x 4 μm, elongated pip-shaped, smooth. Spore print white.

178 *Marasmiellus foetidus = Micromphale foetidum* [= ***Gymnopus foetidus***]

Caps c. 35 mm across, the gill pattern clearly visible through the thin cap tissue as striations and furrows. This fungus has a strong unpleasant odour. Grows on decaying wood mulch. **Spores** c. 8 x 4.5 μm, ellipsoidal, smooth. Spore print white.

177

178

179

180

179 *Marasmius alveolaris*

Caps to 5 mm across, with convex bulges between the widely spaced gills. The slender hair-like stems, up to 12 mm long, are dark brown to black. The species colonizes shed eucalypt bark, where it often forms extensive colonies. Dried caps readily rehydrate with moist conditions, regaining their original size. **Spores** c. 10 x 4.5 µm, ellipsoidal, smooth. Spore print white.

180 *Marasmius elegans*

Caps to 40 mm across, convex to deeply convex, with matt or velvet-like texture. Stems tough, pale above and darker below. A common species found in native forests and introduced pine forests. **Spores** c. 10 x 5 µm, ellipsoidal, smooth. Spore print white.

181 *Marasmius oreades*

Caps to 45 mm across, unevenly convex, translucent-striate with smooth to matt surface. The gills are cream and somewhat distant. It is common in pastures, lawns and nature strips, where it forms 'fairy rings'. It is edible, but the tough stems should be discarded before cooking. **Spores** c. 9 x 5.5 µm, ellipsoidal, smooth. Spore print white.

182 *Melanoleuca melaleuca*

Caps to 80 mm across, convex, becoming unevenly flattened with inturned margin. Smooth, moist, hygrophanous. Gills crowded, whitish to pinkish cream. The barbed, harpoon-like cystidia on the gills are a distinctive microscopic generic character. Associated with pines on calcareous soils. **Spores** c. 8 x 5.5 µm, ellipsoidal, minutely warty, amyloid. Spore print cream.

181

182

183 *Melanoleuca* sp.

Caps to 80 mm across, shallowly funnel-shaped at maturity. This unidentified species is found around forest margins and in grassy clearings. Metuloidal cells are present on the gills, a common character in this genus. **Spores** c. 9 x 5.5 µm, ellipsoidal, finely warty or roughened. Spore print white.

184 *Melanophyllum haematospermum*

Caps to 20 mm across, with pendulous veil remnants on the margin when young. Gills at first scarlet, becoming brown with maturity. A distinctive feature is the spore colour change. Gregarious in native forest and pine forest. **Spores** c. 6.5 x 3.5 µm, finely rough. Spore print dark green when fresh, drying to brown.

185 *Melanotus hepatochrous* [= ***Deconica horizontalis***]

Caps to 20 mm, with short lateral stem. Margin inturned in young caps. A small distinctive fungus that colonizes dead wood and bark. Common. **Spores** c. 7.5 x 4.5 µm, ellipsoidal, smooth, with germ pore. Spore print purple-brown.

183

184

185

186 *Montagnea arenaria*

Caps to 30 mm across, with flattened central disc. Gills hard, brittle and very close, like the leaves of a book, tending to expand outwards and become ragged. The hard woody stem, up to 80 mm high, emerges from a sheathing volva. Resembles a decaying, dry *Coprinus*. Found on desert sands. Uncommon. **Spores** c. 20 x 10 µm, ellipsoidal-elongate, smooth with eccentric germ pore. Spore print black.

187 *Mycena albidocapillaris*

Caps to 5 mm across, with minute central depression, translucent-striate. Stem translucent, very slender with scattered short hairs. Gregarious, often abundant on decaying fern fronds, leaves and twigs in wet forest. **Spores** c. 8.5 x 5 µm, ellipsoidal, smooth. Spore print white.

188 *Mycena albidofusca*

Caps to 20 mm across, convex to broadly conical, translucent-striate, except for the distinctive pale, flattened umbo. A widespread and locally abundant species. Gregarious on litter in eucalypt forests. **Spores** c. 9 x 5.5 µm, ellipsoidal, smooth. Spore print white. (**188b** shows dorsal view)

186

188a

187

188b

189 *Mycena.* aff. *atrata*

Caps to 25 mm across, broadly convex to broadly conical, often with a prominent umbo, minutely radially wrinkled, moist to greasy to touch, dark brown to blackish brown, sometimes with a paler edge. Gills grey to grey-brown. Gregarious to caespitose on decaying eucalypt logs and stumps.
Spores c. 9.5 x 7 µm, broadly ellipsoidal, smooth. Spore print white.

191

192

190 *Mycena austrofilopes*

Caps to 20 mm across, broadly conical with matt surface. Stems long and slender. It is somewhat similar in appearance to *M. cystidiosa* but is paler and shorter. Gregarious, forming colonies in leaf litter, and frequently amongst moss. Common. **Spores** c. 10 x 5.5 µm, ellipsoidal, smooth. Spore print white.

191 *Mycena austrororida* [= ***Roridomyces austrororidus***]

Caps to 10 mm across, occasionally larger, deeply convex, moist and usually speckled with tiny brown dots. Stems covered with thick clear gluten. Gregarious to caespitose on dead logs and branches in eucalypt forest and rainforest and frequently on pine cones. **Spores** c. 11 x 5 µm, ellipsoidal, smooth. Spore print white.

192 *Mycena carmeliana*

Caps to 20 mm across, convex, becoming plano-convex, minutely radially wrinkled, viscid, pale grey-brown with paler margin. Stems slender with numerous short hairs and terminating in a distinctive, striate orange disc. It has an ammonia-like odour. **Spores** c. 7 x 5 µm, ellipsoidal, smooth. Spore print white.

193 *Mycena clarkeana*

Caps to 40 mm, broadly conical to bell-shaped, translucent-striate with grooved margin. Shape and colour fairly distinctive. Widespread, from dry to wet eucalypt forest and rainforest, caespitose to densely caespitose, on stumps, decaying logs and dead standing tree trunks. **Spores** c. 12 x 8 µm, ellipsoidal, smooth. Spore print white.

194b

194 *Mycena cystidiosa*

Caps to 20 mm across, broadly conical, radially striate. Possibly our tallest *Mycena*. Stems have abundant white basal hairs and can be up to 200 mm long. Another distinctive character is the often plentiful hair-like rhizomorphs with minute capitate tips associated with the fungus (**194b**). Common on decaying forest litter. **Spores** variable c. 8 x 5.5 µm, ellipsoidal, smooth. Spore print white

196

195 *Mycena* aff. *epipterygia*

Caps to 15 mm across, conical to broadly conical, translucent-striate and viscid. The yellow stems are viscid to subglutinous. Colour may vary from shades of yellow to pale grey. A cucumber-like odour is produced when bruised. A common fungus found in most forest types, including pine forest. It grows on dead wood, bark, leaf litter and pine cones. **Spores** c. 9 x 6.5 µm, ellipsoidal, smooth. Spore print white.

196 *Mycena fumosa*

Caps to 18 mm across, ovoid at first, becoming broadly convex to almost flattened, translucent-striate, viscid, shiny. Apart from its creamy-brown colour it is very similar to *Mycena interrupta* in form and habitat. Gregarious to caespitose on decaying logs in wet eucalypt forest. **Spores** c. 8 x 5 µm, ellipsoidal to nearly oblong, smooth. Spore print white.

197 *Mycena interrupta*

Caps to 15 mm across, deeply convex to convex, translucent-striate and slightly viscid. Although small, the unique bright blue caps with their dark central discs draw immediate attention to the usually small colonies. The white gills are also edged with blue. Found on dead branches and logs, and occasionally on fallen pine cones. **Spores** c. 8.5 x 6 µm, ellipsoidal, smooth. Spore print white.

197

198 *Mycena kurramulla*

Caps to 20 mm across, convex, translucent-striate, rosy pink. Gills arched, decurrent, with prominent red edge. One of the most distinctive Mycenas. Gregarious to caespitose on decaying eucalypt stumps, logs and branches. **Spores** c. 7 x 4 µm ellipsoidal, smooth. Spore print white.

199 *Mycena kuurkacea* (previously included with *M. sanguinolenta*)

Caps to 15 mm across, sometimes larger, bell-shaped to conical, translucent-striate. The slender stems produce a clear red fluid when cut or broken. The fruit bodies grow on decaying wood and bark in eucalypt forest and rainforest. Large colonies are sometimes seen on old mossy logs. **Spores** c. 10 x 6 µm, ellipsoidal, smooth. Spore print white.

198

199

200

200 *Mycena leaiana* var. *australis*

Caps to 30 mm across but usually less, olive-yellow to orange-green, with orange margin, viscid. Gills broad, pinkish orange edged with bright orange inflated cells. Stems with scattered orange flecks. Found on dead wood in fern gullies and wet forest. **Spores** c. 8 x 4.5 µm, ellipsoidal, smooth. Spore print white.

201 *Mycena maldea*

Caps to 5 mm across, convex with central depression. Smell strongly nitric. Similar to *M. albidocapillaris,* which lacks the strong nitric odour. This fragile species often forms large colonies on twigs, bark, leaf litter and occasionally on fallen pine cones. **Spores** c. 8.5 x 5 µm, ellipsoidal, smooth. Spore print white.

202 *Mycena marangania*

Caps to 20 mm across, broadly convex with dark brown umbo, outer half of radius translucent-striate. A notable microscopic character is the abundance of large cystidia on the face and edge of the gills. Gregarious to caespitose on decaying wood, bark and leaf litter in eucalypt forest. **Spores** c. 11 x 6.5 µm, ellipsoidal, smooth. Spore print white.

203 *Mycena mijoi*

Caps to 25 mm across, convex to shallow-convex with flat to slightly depressed apex. Moist and slightly translucent-striate. Stems glutinous. Similar to *M. subvulgaris* but paler and with the cap apex not as depressed. Gregarious to caespitose on leaf litter and forest debris. **Spores** c. 8 x 4 µm, narrow-ellipsoidal, smooth. Spore print white.

204 *Mycena minya*

Caps to 5 mm across, convex to conical with tiny depression, often frosted with minute granules. Found in troops on living or shed eucalypt bark. **Spores** c. 9 x 6 µm, ellipsoidal, smooth. Spore print white.

203

204

205 *Mycena mulawaestris*

Caps to 25 mm across, conical to broadly conical, translucent-striate, glutinous to very glutinous. Young caps are almost black, becoming medium brown with the apex remaining dark. Caespitose to gregarious on decaying wood in wet eucalypt forest. **Spores** c. 9 x 6 µm, ellipsoidal, occasionally nearly oblong, smooth. Spore print white.

206 *Mycena nargan*

Caps to 30 mm across, conical with rounded apex, dark brown to almost black when young and ornamented with small white evanescent fibrillose scales. Mature specimens become expanded, lose the scales, and are difficult to recognize. Gregarious to slightly caespitose on dead wood. **Spores** c. 9 x 6 µm, ellipsoidal, smooth. Spore print white.

207 *Mycena nivalis*

Caps to 11 mm across, convex with a slight central depression, slightly viscid, translucent-striate. Gregarious on decaying litter in mixed *Eucalyptus/ Leptospermum forest*. Uncommon. **Spores** c. 8 x 4.5 µm, narrowly ellipsoidal, smooth. Spore print white.

206

208a

209

208 *Mycena pura*

Caps to 40 mm across, broadly conical to flattened, smooth, slightly greasy to touch. This pale, translucent fungus comes in white (**208b**) or shades of lilac and pink (**208a**), and suggests the genus *Hygrocybe*. Gregarious in pine and eucalypt forests. Odour radish-like. **Spores** c. 7 x 4 µm, ellipsoidal-cylindrical, smooth. Spore print white.

209 *Mycena subgalericulata*

Caps to 25 mm across, conical with pointed to rounded umbo, caps becoming broader with age. Caespitose to densely caespitose on dead wood, also on fibrous bark of living trees. Common in eucalypt forest and rainforest. **Spores** c. 10.5 x 7.5 µm, ellipsoidal, smooth. Spore print white.

210 *Mycena subvulgaris*

Caps to 20 mm across, convex, with central depression, strongly striate. Gills white, decurrent. Stems white, slender and very glutinous. Gregarious to densely gregarious on leaf litter and forest debris. **Spores** c. 9 x 4.5 µm, narrow-ellipsoidal, smooth. Spore print white.

210

211 *Mycena toyerlaricola*

Caps to 15 mm across, conical with rounded apex, translucent-striate, nearly transparent. Gills red-brown, distant. Stems slender, brownish red, releasing watery red fluid when cut or broken. The species appears to be restricted to *Nothofagus cunninghamii* (myrtle beech) forest, where it is widespread and fairly common. **Spores** c. 8.5 µm long, ellipsoidal, smooth. Spore print white.

211

213

212

212 *Mycena vinacea*

Caps to 40 mm across, convex, becoming flattened, or finally upturned at the rim. This species can be confused with *Mycena pura*, which is usually paler, with white to pink colours. Both species have a radish-like odour. Gregarious on the ground in both native and pine forests. **Spores** c. 8.5 x 4.5 µm, ellipsoidal, smooth. Spore print white.

213 *Mycena viscidocruenta* [= ***Cruentomycena viscidocruenta***]

Caps to 15 mm across, convex with shallow central depression, translucent-striate and viscid when moist. Gill colour similar to the cap or a little paler. The bright red stem is covered with a thick layer of clear gluten. A small, attractive fungus found on dead twigs and leaf litter in a wide range of forest and heathland habitats. **Spores** c. 9 x 4.5 µm, narrowly ellipsoidal, smooth. Spore print white.

214 *Mycena yuulongicola*

Caps to 30 mm across, broadly conical, becoming broadly bell-shaped or somewhat flattened with age. Gregarious to densely caespitose on dead eucalypts and other forest trees. **Spores** c. 9 x 6 µm, ellipsoidal, smooth. Spore print white.

214

215 *Mycena* sp.

Caps to 3 mm across, conical, becoming bell-shaped, striate, matt, with dark apex. It resembles a miniature form of *Mycena subgalericulata* but is microscopically distinctive. Found on decaying mossy wood in moist forest habitats. **Spores** c. 10 x 6 µm, ellipsoidal, smooth. Spore print white.

215

216a

216b

216 *Mycena* sp.

Caps to 30 mm across, broadly convex, slightly viscid when wet, margin translucent-striate. Gills white with red-brown edges. Gregarious in moss beds on sheltered forest floor. **Spores** c. 9.5 x 5 µm, ellipsoidal with slightly flattened sides, smooth. Spore print white.

217 Mycena sp. [= *Clitocybula* sp.]

Caps to 25 mm across, convex, becoming broadly convex with dark, flattened to slightly depressed centre, radially fibrillose with fine dark fibrils. Gills arched, appearing decurrent. Fruit bodies with translucent, watery appearance. Gregarious to caespitose, sometimes in large colonies on tree trunks and dead wood in wet forest. An uncommon and somewhat atypical *Mycena*. **Spores** c. 6 µm diam., globose, smooth. Spore print white.

218 *Mycena* sp.

This photo shows an underview of part of a colony of an unidentified species of *Mycena*. The gills are very unequal, the longest being attached decurrently to the stem. The colony shown was growing on a decaying log bridging a mountain stream. **Spores** not examined. Spore print white.

219 *Mycena* sp.

Caps to 3 mm across, convex, translucent-striate above with surface bulbous between gill lines. Gills few, connected to a collar through which the stem passes. Colour from blue to almost white. Occurs on dead fern rachises in wet forest. It is luminescent and has a strongly nitric odour. **Spores** c. 8 x 5.5 µm, broadly ellipsoidal to almond-shaped. Spore print white.

220 *Hemimycena* sp.

Caps to 15 mm across, shallow-convex, smooth, the gills shallow, decurrent in some populations. All white at first and becoming yellow with age or when handled. On decaying wood in moist forest. Widespread but uncommon. **Spores** c. 10.5 x 5.5 µm, ellipsoidal, smooth. Spore print white.

219

221 *Neolentinus dactyloides*

Caps to 90 mm across, broadly convex to nearly flat, densely hairy with coarse bristle-like hairs. The tough, leathery fruit bodies arise from buried pseudosclerotia (**221b**). The gills are close, often forked and unevenly toothed. Widespread in eucalypt forest but fruits freely only after fire. This species is a member of the woody-pore group but is included here because of the presence of gills. **Spores** c. 11 x 6 µm, ellipsoidal, smooth. Spore print white.

222 *Omphalina chromacea* [= ***Lichenomphalia chromacea***]

Caps to 15 mm across, depressed to deeply funnel-shaped. A lichenised fungus found growing on algae-covered earth, including bush tracks, old rabbit mounds, earth banks, etc. Its bright chrome yellow colour is distinctive. **Spores** c. 8 x 5 µm, ellipsoidal, smooth. Spore print white.

221a

221b

223 *Omphalina umbellifera* [= **Lichenomphalia umbellifera**]

Caps to 15 mm across, resembling *O. chromacea*, but usually smaller, more deeply funnel-shaped and light brown to yellow-brown in colour. It is a lichenised fungus occurring in similar habitats to *O. chromacea*. **Spores** c. 8 x 5 µm, ellipsoidal, smooth. Spore print white.

224 *Omphalina* sp. [= ***Lichenomphalia sp.***]

Caps variable up to 20 mm across. This species also resembles *O. chromacea* (**222**) but is larger, has fewer gills, and is all white. It appears in numbers in arid and semi-desert country on apparently bare ground after soaking rain. **Spores** c. 7.5 x 4.5 µm, ellipsoidal, smooth. Spore print white.

225 *Panaeolina foenisecii*

Caps to 20 mm across, convex to broadly conical, strongly hygrophanous. Gills at first grey-brown, becoming dark brown with pale margin. Occurs in grassy paddocks and lawns, especially after mowing. **Spores** c. 14 x 7.5 µm, variable, ellipsoidal with irregular dark brown warts. Spore print black.

226 *Panaeolus papilionaceus*

Caps to 40 mm across, parabolic to bell-shaped, surface shiny grey-brown and frequently developing a pattern of cracks on drying. Gills dark, mottled, becoming black with pale edges when mature. Gregarious on herbivore dung, especially of horses and cattle. **Spores** c. 16 x 9 µm, variable, ellipsoidal, smooth, with germ pore. Spore print black.

224

225

226

227 *Panaeolus sphinctrinus*

Caps to 35 mm across, parabolic to campanulate, usually found as gregarious colonies on horse and cattle manure. Immature caps are semiglobose; the dark gills are mottled with grey patches. **Spores** c. 10 x 9 µm, ellipsoidal, smooth, with germ-pore. Spore print black.

228 *Panellus ligulatus*

Caps to c. 40 mm across, 50 mm long, spathulate with lateral stem. A distinctive species found on rotting wood, often on decaying eucalypt windrows in pine forest. **Spores** c. 8 x 3.5 µm, subcylindrical, smooth. Spore print white.

227

229 *Dictyopanus pusillus* [= ***Panellus pusillus***]

Caps to 15 mm across, somewhat kidney-shaped with tiny lateral stems. The under surface has pores with c. 3 pores per mm. It is closely related to the gilled fungus *Panellus stipticus* but is smaller. Both species grow on dead wood, including log barriers in bush car parks and picnic areas, often forming large colonies. Both are luminescent in darkness. **Spores** c. 4.5 x 2.5 µm, ovoid, smooth. Spore print white.

230

231

232

233

230 *Panellus stipticus*

Caps to 25 mm across, sometimes larger, mostly smooth when fresh, forming almost stemless, soft, kidney-shaped fruit bodies on decaying logs and stumps. The gills, which may be sparsely joined by cross-members, exude small quantities of milky latex that is sticky to touch, a useful field character. It is luminescent in darkness like its close relative, *Panellus pusillus*. **Spores** c. 5 x 3 µm, ellipsoidal, smooth. Spore print white.

231 *Panus fasciatus*

Caps to 50 mm across, funnel-shaped with inrolled margins, the surface covered with dense, stiff hairs. Texture thin and leather-like, becoming hard and brittle when dry but rehydrating after rain. The purplish brown gills are decurrent and occasionally forked. Found on dead wood in dry regions including eucalypt and mallee woodland and desert habitats. **Spores** c. 7.5 x 4 µm, ellipsoidal, smooth. Spore print white.

232 *Panus* sp.

Caps to 50 mm across, irregularly shallowly funnel-shaped with inrolled margin, matt. Young caps with small dark flecks. Gills close, shallow, decurrent and tending to fork near the ends. Found on decaying wood in montane forest. Uncommon. **Spores** c. 7.5 x 4 µm, ellipsoidal, smooth. Spore print white.

233 *Phaeocollybia tasmanica*

Caps to 60 mm across, broadly conical with shallow umbo. Gills close, lilac at first, becoming brown as the spores mature. The stem is slightly swollen at ground level, and tapers downwards to a long root-like tail, probably attached to buried decaying wood. An uncommon species of *Nothofagus cunninghamii* (myrtle beech) forest. **Spores** c. 6 x 4 µm, ellipsoidal with small warts. Spore print rust-brown.

236

234 *Phaeocollybia* sp.

Caps to 55 mm across, convex, umbonate, becoming broadly convex with low umbo, viscid when moist. Gills close, almost free from the stem, pale at first, becoming lilac, then brown with spores. Stem width even and deeply rooting to buried decaying wood. Associated with myrtle beech in cool temperate rainforest. **Spores** c. 4.8 x 4 µm, broadly almond-shaped to ovoid with flattened side, warty. Spore print rust-brown.

235 *Pholiota communis*

Caps to 75 mm across, broadly convex to flattened and slightly umbonate, viscid when moist, the centre with dense fibrillose scales that become separated and paler towards the margin. Stems brown, fibrillose, becoming fibrillose scaly towards the base. Gregarious to caespitose in eucalypt and pine forest among forest litter. **Spores** c. 8.5 x 5.5 µm, ellipsoidal with germ pore, smooth. Spore print dull brown.

236 *Pholiota highlandensis* (previously known as *Pholiota carbonaria*)

Caps to 30 mm across, deeply convex, slightly viscid when young, becoming broadly convex or expanded with age. The yellowish brown stems are covered with small brown to red-brown squamules. Gregarious to clumped on burnt ground amongst charcoal. Common after bush fires. **Spores** c. 7 x 4.5 µm, ellipsoidal, with germ pore, smooth. Spore print brown.

237 *Pholiota malicola*

Caps to 60 mm across, convex, becoming flatter with age, hygrophanous. The creamy yellow gills become darker as the spores mature. Caespitose on stumps and buried dead wood such as decaying tree roots, etc. **Spores** c. 10 x 5 µm, ellipsoidal, smooth, with germ pore. Spore print chocolate brown.

238 *Pholiota squarrosipes*

Caps to 70 mm across, convex, becoming broadly convex, sometimes slightly umbonate with fibrillose scales; viscid when moist; colour variable from brownish cream to bright tan. Stems with shaggy fibrillose scales below the membranous annulus. Caespitose on the ground in native forests and pine forests. **Spores** c. 7.5 x 4.5 µm, ellipsoidal, smooth, with distinct germ pore. Spore print dull brown.

239 *239a Pleurotopsis longinqua = Panellus longinquus* [= ***Scytinotus longinquus***]

Caps to 60 mm wide and 40 mm radius, attached by a short lateral stem. The beautifully translucent caps range from white to pink, with a clear gelatinous layer below the cap surface. It grows on fallen, or more often standing, small, dead trunks in wet forest. **239b** shows lower surface. **Spores** c. 8 x 4 µm, ellipsoidal, curved like jelly-beans, smooth. Spore print white.

240 *Pleurotus purpureo-olivaceus*

Caps to 100 mm long and 80 mm across with lateral stems. The dark olive to purple-olive coloured cap surface and grey gills (**240b**) are distinctive. Gregarious on dead logs and branches in wet forest and rainforest of eastern Australia. **Spores** c. 7 x 3 µm, ovoid to ellipsoidal, smooth. Spore print white.

241 *Pluteus atromarginatus*

Caps to 90 mm across, broadly convex to flattened, with dark grey radial fibrils. Gills white to pale pink with dark to almost black edges. One of our larger species of *Pluteus*. Found on decaying wood. **Spores** c. 7 x 4.5 µm, ellipsoidal, smooth. Spore print pink.

242 *Pluteus cervinus*

Caps to 100 mm across, convex at first, becoming broadly convex and finally flattened to undulate, covered with fine dark scales. Gills close, whitish at first, becoming pink with maturity. Metuloidal cystidea of gills terminating in several prongs or horns. Frequent on decaying wood, specially woodchips and sawdust. **Spores** c. 7.5 x 5.5 µm, broadly ellipsoidal, smooth. Spore print pink.

241

242

243 *Pluteus lutescens*

Caps to c. 40 mm across, with matt to finely granulose surface. A fragile species found on decaying wood. Pale buff gills become pink as the spores mature. **Spores** c. 7 x 6 µm, subglobose, smooth. Spore print pink.

244 *Psathyrella asperospora* [= *Lacrymaria asperospora*]

Caps to 100 mm across, conical to convex, flattened with age, surface covered with radial fibrils. Gills dark grey, becoming mottled when mature with pendulous droplets of black spore-containing fluid, giving rise to the former generic name, literally 'Weeping Mary'. A common fungus of forest edges where it often forms dense caespitose colonies. **Spores** c. 10 x 7 µm, ellipsoidal, coarsely warty (mulberry-like). Spore print black.

243

244

245 *Psathyrella echinata*

Caps to 15 mm across. Young caps are densely covered with soft projecting fibrillose spines, which soon fall off. Densely gregarious on dead wood and bark. A short-lived species. **Spores** c. 6.5 x 4 µm, ellipsoidal, smooth, with germ pore. Spore print dark brown.

246 *Psathyrella* aff. *pennata*

Caps to 25 mm across, semiglobose, becoming parabolic, radially fibrillose with inrolled margin. At first covered with dense white tomentum that breaks up into an ornamental fringe on the cap margins of young specimens. The fringe falls away with age. Fruit bodies grow from buried decaying wood and decaying logs, etc. **Spores** c. 7.5 x 5.5 µm, ellipsoidal, smooth. Spore print almost black.

247 *Psilocybe subaeruginosa*

Caps to 70 mm across, at first conical with incurved margin, becoming broadly convex or with rounded to acute umbo. Cap texture smooth with greasy or waxy feel. Young cap rims are connected to the stem by a white cobweb like veil. The slender, somewhat flexuose stem is firm and fibrous. All parts stain blue with age, or when handled or bruised. Gregarious in open forest, amongst grass, on decaying wood, stumps and wood chips and sometimes on tree-fern trunks in wet forest. **Spores** c. 12 x 7 µm, ellipsoidal, smooth with germ pore. Spore print purple-black.

248 *Resupinatus applicatus*

Caps to 15 mm across, inverted, pendulous, with short stem. Cap surface at first tomentose, becoming wrinkled, gelatinous below surface with distinctive branched, proliferating hyphae. Common in pine forests on dead branches and twigs, occasionally on eucalypt debris. **Spores** c. 5 µm diam., subglobose, smooth. Spore print white.

247

248

249

250

249 *Resupinatus cinerascens*

Caps to c. 12 mm across, clothed with hoary tomentum that becomes less obvious when mature. Fruit bodies darken with age. The flesh is gelatinous and translucent. A small but beautiful species found on decaying wood and woody bark. **Spores** c. 7 x 4.5 µm, ellipsoidal, smooth. Spore print white.

250 *Rickenella fibula*

Caps to 10 mm across, convex, with a distinct central depression. The slender stem has scattered short protruding hairs. It grows amongst moss colonies, the stem length depending on the depth of the moss bed. **Spores** c. 6 x 3 µm, ellipsoidal, smooth. Spore print white.

251 *Rickenella setipes* [= ***Rickenella swartzii***]

Caps to 12 mm across, funnel-shaped, on long tapering stems. Grows in grass or moss, frequently along forest margins. Stem length varies with thickness of grass or moss beds. **Spores** c. 4.5 x 2.5 µm, ellipsoidal, smooth. Spore print very pale yellow.

252 *Rozites foetens* [= ***Cortinarius perfoetens***]

Caps to 100 mm across, the margin incurved with cog-like remnants of the universal veil at intervals around the rim. Stems unusually furry or fibrillose, with a fibrillose membranous ring. This fungus has a strong, distinctive, unpleasant odour with a suggestion of hot or burnt rubber. Associated with *Nothofagus cunninghamii* (myrtle beech) in rainforest. **Spores** c. 10.5 x 6.5 µm, broadly almond-shaped, warty. Spore print rust-brown.

251

252

253 *Rozites metallica* **[= *Cortinarius metallicus*]**

Caps to 100 mm across, at first viscid with white flecks of veil remnants that fall away as the caps expand. The initial blue-grey colour fades to pale grey and finally to yellow-brown. Associated with *Nothofagus cunninghamii* and, in Tasmania, also with *Nothofagus gunnii*. Uncommon. **Spores** c. 10.5 x 8 µm, broadly almond-shaped, warty. Spore print rust-brown.

254 *Rozites roseolilacina* **[= *Cortinarius roseolilacinus*]**

Caps to 150 mm across, convex at first, becoming flattened and undulate with age. Immature caps are covered with a thin powdery veil, which is visible as a pendulous remnant on the rim. Colour fades to brownish yellow tones with age. Associated with *Eucalyptus* and *Leptospermum* forest. **Spores** c. 10.5 x 7.5 µm, broadly almond-shaped, warty. Spore print rust-brown.

255 *Russula flocktoniae*

Caps to 100 mm or more across, broadly and sometimes unevenly convex, becoming centrally depressed and often funnel-shaped. Cap surface matt with velvet appearance. Gills white and somewhat distant. The brittle flesh has a peppery taste. Common in eucalypt forest and heathland. **Spores** c. 9 x 7 µm, subglobose with amyloid reticulations and warts. Spore print cream.

256 *Russula kalimna*

Caps to 70 mm across, convex, becoming flattened and slightly depressed when mature. The cap colour shows various tones of pale blue and yellow. The cap surface is often finely cracked. The stem is unusually short, barely lifting the cap above the ground, giving the fungus a chunky appearance. Solitary or scattered in eucalypt forest. Widespread but uncommon. **Spores** c. 7 x 5 µm, subglobose, ornamented with low ridges and warts. Spore print pale yellow.

255

256

257 *Russula lenkunya* (formerly included in *Russula mariae*)

Caps to 100 mm or more across, broadly convex with central depression, becoming flattened; surface matt. The white gills are sometimes finely toothed with edges often tinged red or pink. Common in eucalypt forest. **Spores** c. 9 x 8 µm, subglobose, with fine amyloid warts. Spore print cream.

258 *Russula persanguinea*

Caps to 90 mm across, convex with low central depression and faintly striate margin when mature; slightly viscid when moist. May be confused with other red species. Gregarious in eucalypt and mixed forest. **Spores** c. 9 x 8 µm, subglobose, with reticulate, amyloid ornamentation. Spore print white.

259 *Russula purpureoflava*

Caps to 60 mm across, often centrally depressed, reddish purple and slightly viscid when moist. Gills yellow. Common in eucalypt forest and heathland. **Spores** c. 9 x 7.5 µm, subglobose, with amyloid ridges and warts. Spore print cream.

257

258

259

260 *Russula rosea*

Caps to 100 mm across, convex, becoming broadly convex; surface matt. Stem colour similar to cap but becoming paler with age. Occurs in tall eucalypt forest. Uncommon. **Spores** c. 9 x 9 µm, globose, with large amyloid reticulations. Spore print cream.

261 *Schizophyllum commune*

Caps to 30 mm across; irregularly fan-shaped with lateral stem; densely hairy above, pale pink to creamy pink, becoming greyish with age. A distinctive character is the split, or double-edged, gills (**261b**). It is a cosmopolitan fungu that colonises many kinds of dead wood. **Spores** c. 6.5 x 2.5 µm, cylindrical or slightly curved, smooth. Spore print white to pale orange.

262 *Setchelliogaster tenuipes*

Fruit bodies up to 30 mm high and 15 mm diameter, with a web-like cortina below. Caps do not expand fully. Gills are irregular, almost enclosed, with cross connections. It grows with *Eucalyptus* species and is closely related to the gilled genus *Descolea*. Uncommon. **Spores** c. 18 x 10 µm, ellipsoidal, warty. Spores brown in mass.

261a

261b

262

263 *Simocybe phlebophora*

Caps to 30 mm across, convex, becoming broadly convex, translucent-striate. The cap surface has a distinctly wrinkled, gelatinous texture that is raised, thickened and more wrinkled over the centre. It is found on decaying stumps and logs in eucalypt forest and rainforest. Widespread. **Spores** not examined. Spore print brown.

264 *Stropharia semiglobata*

Caps to 40 mm across, nearly hemispherical, becoming slightly expanded with age; surface smooth, viscid. The broad gills are pale at first, becoming progressively darker with maturing spores, finally black. It grows singly, or in small colonies on herbivore dung. The variable size of the fruit bodies relates to the available food reserves of the substratum. Common. **Spores** c. 18 x 10 μm, ellipsoidal, smooth, with germ pore. Spore print purple-black.

265 *Tephrocybe* aff. *rancida*

Caps to 40 mm across, dark brown to almost black with grey gills. Smell rancid. Found on the ground after fire where it grows singly or in small clusters. **Spores** c. 7 x 4 μm, ellipsoidal, smooth. Spore print white.

265

266

266 *Tricholoma austrocolossum*

Caps to 80 mm across, at first unevenly convex becoming irregularly flattened, smooth, fleshy. The pinkish cream gills become mottled with age. The robust stem has an annulus, an uncommon character in the genus. It is found on the ground in eucalypt forest. **Spores** c. 7.5 x 6 µm, ovoid, smooth. Spore print white.

267 *Tricholoma eucalypticum*

Caps to 150 mm across, irregularly convex, becoming flattened to undulate and distorted owing to crowding, slightly viscid when moist, finely fibrillose to smooth when dry. Gills at first pale becoming mottled brown with age. Flesh bitter to taste. Caespitose under eucalypts. **Spores** c. 6 x 4.5 µm, ellipsoidal, smooth. Spore print white.

268 *Tricholoma virgatum*

Caps to 80 mm across, conical at first, becoming broadly convex with low umbo; the radially fibrillose surface usually splits to expose the paler cap tissue. Usually found in pine forest. **Spores** c. 6.5 x 5 µm broadly ellipsoidal, smooth. Spore print white.

269 *Tricholomopsis rutilans*

Caps to 150 mm across, convex, becoming broadly convex to flattened or undulate, the surface yellow, covered with purplish red fibrils. Gills yellow. Stems yellow, under a cover of purplish red fibrils. This spectacular fungus is found on or near decaying stumps and wood chips, especially of pines. **Spores** c. 7.5 x 5.5 µm, ellipsoidal, smooth. Spore print white.

267

268

269

270 *Trogia straminea*

Caps to 30 mm across, translucent-striate, usually deeply funnel-shaped. The pale yellow-brown decurrent gills are cross-connected with shallow veins. A distinctive species found on dead wood in wet forest and rainforest. **Spores** not examined. Spore print white.

271 *Tubaria rufofulva*

Caps to 45 mm across, conical to convex, with scattered fibrillose scales, becoming flattened with central umbo. The scurfy scales and membranous ring present on the stem of young fruit bodies disappear with age. A widespread fungus found on decaying wood and forest debris in moist habitats. **Spores** c. 8.5 x 5.5 µm, ellipsoidal, smooth. Spore print rust-brown.

270

272 *Volvariella speciosa* [= ***Volvopluteus gloiocephalus***]

Caps to 100 mm across, occasionally larger, conical, becoming broadly conical and finally plane with central umbo, smooth and viscid. Gills pink with maturing spores. Stem base surrounded by a large ruptured volva that originally enclosed the developing fruit body. Found in soil rich in organic matter, lawns and roadsides. **Spores** c. 16 x 10 µm, ellipsoidal, smooth. Spore print brownish pink. The darker **272b** is sometimes distinguished as **var. *gloiocephala***.

273

274

273 *Xeromphalina leonina*

Caps to 5 mm across, nearly hemispherical and translucent-striate. Although the tiny caps of this fungus are only a few millimetres across, they frequently form dense colonies. The preferred habitat is decaying logs. It is sometimes found on eucalypt remains in pine forest. **Spores** c. 6.5 x 4 µm, ellipsoidal, smooth. Spore print white.

274 *Xerula gigaspora = Oudemansiella radicata var. australis*
[= ***Oudemansiella gigaspora***] Rooting shank

Caps to 80 mm across, broadly convex to almost flat, viscid, smooth. The common name refers to the long, tapering root-like extension of the stem. It is easily recognized by its very long white stem and viscid, brown cap with pure white gills. A common fungus found in grasslands, native forest and in pine plantations. **Spores** c. 13 x 10 µm, ellipsoidal, smooth. Spore print white.

1.2 FUNGI WITH FORKED GILLS: *Paxillus* and allies

These fungi are closely related to the fleshy pore-fungi, but have forked gills instead of pores. The gill tissue can easily be separated from the cap tissue. Most members of this group change colour, usually to green or brown, when bruised,

275 *Austropaxillus infundibuliformis*

Caps to 150 mm across, usually deeply funnel-shaped at maturity, sometimes eccentric or lobed. Gills obviously forked. This large fungus is common along roadsides in eucalypt forest. It is usually gregarious, but occasionally caespitose. **Spores** c. 15 x 6 µm, long-ellipsoidal, smooth. Spore print bright rust-brown.

276 *Hygrophoropsis aurantiaca*

Caps to 60 mm across, funnel-shaped with inrolled margin. Usually found in moist upland forest and rainforest. Resembles *Paxillus* or *Cantharellus*. **Spores** c. 6 x 4 µm, ellipsoidal, smooth. Spore print pale yellow.

275

276

277a

277b

277 *Omphalotus nidiformis*

Caps to 200 mm across, deeply funnel-shaped with central to lateral stems, sometimes fan-shaped. Cap colour varies from white to shades of brown or purple, the central region often being deeply coloured. This fungus is best known for its bright luminescence at night. It colonises stumps and logs in eucalypt and pine forest and is occasionally seen at the base of living trees, in which it produces a white heart-rot. POISONOUS, often confused with edible species. **277b** *Omphalotus nidiformis* photographed by its own light. **Spores** c. 8 x 6 µm, ellipsoidal, smooth. Spore print white.

278 *Paxillus involutus*

Caps to 150 mm across, at first convex, becoming flattened or unevenly funnel-shaped, the margins persistently inrolled. Cap surface frequently developing patterns of cracking or radial or concentric zonation. An introduced species from the northern hemisphere and frequently found under birches, oaks, conifers and other exotic trees. **Spores** c. 8 x 5 µm, ellipsoidal, smooth. Spore print yellowish brown.

279 *Phylloporus clelandii*

Caps to 200 mm across, deeply convex at first, becoming broadly convex, finally unevenly flattened and undulate. The broad yellow to greenish yellow gills and the stem's flesh slowly turn green when bruised or broken. Usually found under eucalypts, but not common. **Spores** c.12 x 7.5 µm, ellipsoidal, smooth.

281

280 *Phylloporus rhodoxanthus*

Caps to 120 mm across, broadly convex at first, becoming flattened to undulate or slightly depressed. Gills broad, often forked, with cross veins near stem, yellow to greenish yellow, staining green when bruised. Gregarious in eucalypt forest and woodland. **Spores** c. 13 x 5.5 µm, ellipsoidal, smooth. Spore print brown.

281 *Meiorganum curtisii* [= ***Pseudomerulius curtisii***]

Fruit bodies to 60 mm across, sometimes fused together to form larger fruiting masses. When laterally attached the fungus forms shelves on the sides of decaying logs. Often broadly attached on the underside of decaying logs of native and introduced trees. **Spores** c. 4 x 1.5 µm, ellipsoidal, smooth. Spore print yellow-brown.

282 *Tapinella panuoides*

Fruit bodies form fan-shaped lobes to 150 mm across and 200 mm long, often occurring as dense, overlapping colonies on decaying tree stumps and logs, particularly pine stumps and also on pine chips and sawdust. Lilac mycelium is found at the base of the lateral stems. **Spores** c. 5 x 3.5 µm, ellipsoidal, smooth. Spore print yellow-brown.

282

1.3 FLESHY-PORE FUNGI: *Boletus* and allies

The species in this group are commonly referred to as Boletes. They are related to the gilled fungi, but the spore-producing tissue lines the inside of the pores. The pore layer is easily separated from the cap tissue. The caps are usually large, with a centrally located stem and the flesh of many species changes colour rapidly when broken.

283

283 *Austroboletus lacunosus [= Austroboletus cookei]*

Caps to 150 mm, but variable, dry with granulose texture, the stem lacunose and not glutinous. Pores pink. Frequent in eucalypt and mixed forest. **Spores** c. 15 x 8 µm, ellipsoidal, warty in central region. Spore print pale brown.

284 *Austroboletus niveus*

Caps to 150 mm across, white and glutinous when young, becoming pale ochraceus with maturity. The lacunose stems are glutinous to viscid and are intensely bitter, a distinctive character. Found in wet to dry forest and woodland. **Spores** c. 18 x 6 µm, long-ellipsoidal, smooth. Spore print pinkish brown.

285 *Austroboletus novaezelandiae*

Caps to 70 mm across with finely granular, or sometimes tessellated, surface. Stems slender and lacunose like other species in the genus. A slender rainforest species with relatively large pores. Usually found in moss beds on rainforest floor. **Spores** c. 20 x 10.5 µm, fusiform, finely warty. Spore print pale brown.

286 *Boletellus emodensis*

Caps to 100 mm across covered with thick, felty, fibrillose scales, a red ground colour showing through as the cap expands. Pore surface at first covered by a felty membrane which separates as the cap expands. The exposed yellow pores rapidly turn dark grey-blue when bruised or broken. Common in eucalypt and mixed forest. **Spores** c. 22 x 10 µm, cylindrical, ridged. Spore print brown.

287 *Boletellus obscurecoccineus*

Caps to 70 mm across, nearly hemispherical at first, becoming broadly convex. Pore surface bright yellow and barely changing when bruised. Stems relatively long compared with other Boletes. Widespread in eucalypt forest and woodland, but not common. **Spores** c. 15 x 6 µm, ellipsoidal with longitudinal ridges. Spore print yellow-brown.

286

288

287

288 *Boletus barragensis*

Caps to 60 mm across, deeply convex at first, becoming broadly convex. Pore surface dark reddish brown and forming a deep depression around the stem. Pores tiny, the pore tissue bruising bluish green. Stems usually expanding downwards, with a reticulated surface. Solitary to gregarious in open *Eucalyptus/Leptospermum* forest. **Spores** c. 18 x 6 µm, nearly cylindrical with tapering ends, smooth. Spore print pale brown.

289

290

289 Bolete (genus unknown)

Caps to 140 mm across, flattened to slightly undulate, almost smooth. Pores pale yellowish cream. Stems c. 40 mm at widest, lacunose below the pores, smoother towards the base. Found in dry to moist eucalypt forest. Uncommon. **Spores** c. 10.5 x 5 µm, broadly fusiform, smooth. Spore print pale brown.

290 *Chalciporus piperatus*

Caps to 80 mm across, hemispherical at first, becoming unevenly flattened, slightly viscid when young. Pores cinnamon to reddish rust, c. 1–2 per mm. Taste hot, peppery. An introduced species usually associated with conifers but becoming naturalised in *Nothofagus cunninghamii* (myrtle beech) forest, where it may become an ecological threat by replacing native fungi. **Spores** c. 10 x 4 µm, ellipsoidal, smooth. Spore print cigar brown.

291 *Fistulinella mollis*

Caps to 90 mm across, deeply convex at first, becoming broadly convex and finally plane or depressed. The pink pore tissue on the underside of this fungus is remarkably soft in texture, collapsing to a custard-like consistency when handled or bruised. Frequent in eucalypt forest. **Spores** c. 18 x 5 µm, cylindrical, smooth. Spore print brown.

292 *Leccinum scabrum*

Caps to 120 mm across, convex, becoming broadly convex to flattened, the surface slightly viscid when moist. Pores whitish at first, becoming dingy brown with age. Stems ornamented with small dark squamulose scales, distinctive for the genus. Introduced from the northern hemisphere and associated with birch trees. **Spores** c. 18 x 6 µm, slender-ellipsoidal, smooth. Spore print brown.

293 *Phlebopus marginatus*

Caps to 1 metre across, probably Australia's largest terrestrial fungus. A weight of 29 kg was recorded for one specimen from western Victoria. The soft, fleshy tissue is a favourite breeding ground for fungus flies, the maggots of which cause rapid putrefaction of the fungus. Fruit bodies solitary to gregarious, sometimes forming rings around *Eucalyptus* trees. It may fruit at any time of the year after rain. Widespread in eucalypt forest. **Spores** c. 8 x 6 µm, ovoid to pear-shaped, smooth. Spore print yellowish brown.

294 *Pulveroboletus* aff. *ravenelii*

Caps to 100 mm across, convex, tan at first, developing a creamy yellow base colour with scattered tan scales when expanded. Annulus tearing unevenly from cap margin. Pores mustard yellow, staining bluish green where bruised. Flesh yellow. Occasional in eucalypt forest. Uncommon. **Spores** c. 10.5 x 5.5 µm, ellipsoidal, smooth. Spore print yellow-brown.

295 *Strobilomyces* sp.

Caps to 110 mm across, convex at first, becoming broadly convex with inturned to pendulous margin, covered with large, densely fibrillose scales resembling a coarse pinecone. Pores pale at first, rapidly staining dark reddish brown when damaged or bruised. Found in wet eucalypt forest and rainforest. **Spores** c. 11 x 7.5 µm, broadly ellipsoidal, with net-like, large, irregular ridges. Spore print blackish brown.

296 *Strobilomyces* sp.

Caps to 70 mm across, remaining deeply convex, the surface felty-fibrillose, breaking into scale-like tessellations as the cap expands. Pores large, pinkish grey, turning orange-brown when bruised. This fungus is less massive than the previous species, the stem relatively slender and the pores larger. It occurs in dry eucalypt forest. Uncommon. **Spores** not examined.

294

295

296

297 *Suillus amabilis*

Caps to 150 mm across, sometimes with veil remnants around the margin, the surface viscid beneath the reddish brown scales. Pores large, angular. Introduced, found under Douglas fir trees. **Spores** c. 10 x 3.5 µm, long-ellipsoidal, smooth. Spore print yellow-brown.

298 *Suillus granulatus*

Caps to 100 mm across, hemispherical at first, becoming broadly convex. The viscid cap skin can easily be peeled off. Pores at first cream, becoming yellow. The pale yellow stems usually develop small brown granular dots. Gregarious to caespitose in pine plantations. **Spores** c. 9 x 3.5 µm, long-ellipsoidal, smooth. Spore print orange-brown.

298

299 *Suillus luteus* Slippery Jack

Caps to 150 mm across, convex, becoming flattened to depressed, glutinous, especially when young, viscid when older, the viscid layer peeling easily from the cap surface. Pores yellow. Annulus forming an irregular purple band around the stem. Gregarious, associated with pines. **Spores** c. 9 x 4 µm, ellipsoidal, smooth. Spore print yellow-brown.

1.4 CORAL FUNGI

The coral fungi range in form from simple or branched clubs to large, complex coral-like structures. Most species grow on the ground, a few on decaying wood. The fertile tissue covers all but the stem. *Clavaria* and allies have white spores. In the *Ramaria* group the spores are yellow-brown.

1.4.1 *Clavaria* and allies

300 *Aphelaria* sp.

Fruit bodies to 200 mm high by 300 mm wide, densely branched, with the branches somewhat flattened. Flesh firm and pliable. In wet eucalypt and pine forest, uncommon. A close-up of a developing fruit body is shown in **300b**. **Spores** c. 6.5 x 5.5 µm, subglobose, smooth. Spore print white.

301 *Clavaria alboglobospora*

Fruit bodies to c. 100 mm high, usually simple, and often growing in dense clumps. Similar to *Clavaria vermicularis* of the Northern Hemisphere. In eucalypt forest and heathland. Uncommon. **Spores** c. 8.5 x 8 µm, globose to subglobose, smooth. Spore print white.

300b

301

302

302 *Clavaria amoena*

Fruit bodies to 100 mm high, simple, cylindrical or somewhat flattened or grooved. Single, gregarious or occasionally caespitose, amongst mosses in woodland, eucalypt forest and rainforest. Common and widespread. **Spores** c. 6.5 x 4.5 µm, ellipsoidal to ovoid, smooth. Spore print white.

303 *Clavaria corallinorosacea* (also known as *Clavulinopsis corallinorosacea*)

Fruit bodies to 50 mm high, simple, mostly cylindrical. The fertile upper part of the club is covered with a powdery bloom of spores. The stem is translucent. Gregarious amongst moss in various forest habitats. **Spores** c. 7 x 4 µm, ellipsoidal to almond-shaped, smooth. Spore print white.

304 *Clavaria miniata* (also known as *Clavulinopsis miniata*) Flame Fungus

Fruit bodies to 100 mm high, simple or sparsely branched, flattened to broadly flattened or grooved, solitary to caespitose. Colour ranges from pink to bright red. Widespread and fairly common across a range of forest habitats. **Spores** c. 8 x 4 µm, subglobose, smooth. Spore print white.

305 *Clavaria zollingeri*

Fruit bodies to 60 mm high, fragile, often finely grooved. Usually found amongst decaying litter under tree-ferns. It is particularly common in Tasmania. **Spores** c. 6 x 5 µm, broadly ellipsoidal, smooth. Spore print white.

306 *Artomyces colensoi*

Fruit bodies to 30 mm high. A delicate coral fungus found on decaying logs in wet forest. It has a similar peppery taste to *C. piperata* (**307**), but is much smaller with different branching. **Spores** c. 5 x 3 µm, ellipsoidal, minutely rough. Spore print white.

307 *Artomyces piperatus* [= ***Artomyces austropiperatus***] Peppery Coral Fungus

Fruit bodies to 100 mm high, rather variable, but easily recognised by its distinctive branching pattern and habit. It has a peppery taste similar to *C. colensoi* (**306**). Found on dead wood in moist forest. **Spores** c. 6 x 4.5 µm, broadly ellipsoidal, minutely roughened. Spore print white.

308 *Clavulina cinerea*

Fruit bodies to 90 mm high, the branches irregular with blunt tips. Fruit bodies vary from single clubs to large, many-branched forms. The smoky-grey colour is often tinged with purple tones. Found on the ground in eucalypt forest, where fairly common. **Spores** variable, c. 8 x 7.5 µm, subglobose, smooth. Spore print white.

307

309 *Clavulina cristata*

Fruit bodies up to 80 mm high, the branch tips frilled to variously toothed, white, discolouring yellowish with age. A variable species. Individual specimens range from almost simple to many-branched. Found on the ground in forest, woodland and heathland. **Spores** c. 10 x 9 µm, subglobse, smooth. Spore print white.

310 *Clavulina subrugosa*

Fruit bodies up to 80 mm high, simple or sparsely branched, variable in form, size and colour and probably part of a complex of species. Widespread in eucalypt forest and woodland. **Spores** c. 11 x 9 µm, broadly ellipsoidal to subglobose, smooth. Spore print white.

311 *Clavulina vinaceocervina*

Fruit bodies up to 60 mm high, usually pinkish brown to brownish mauve with irregular branching. The surface appears powdery when mature because of protruding cystidia and trapped spores. It usually grows amongst moss in eucalypt forest and rainforest. **Spores** c. 9 x 10 µm, globose, smooth. Spore print white.

312 *Macrotyphula juncea*

Fruit bodies up to 100 mm high, simple. White at first, becoming clay-coloured with age. The clubs may occur singly, or as gregarious colonies, arising from small mycelial mats on dead leaves or twigs of eucalypt and other broad-leaved tree species in moist to wet forest. **Spores** c. 8 x 4.5 µm, broadly cylindrical, smooth. Spore print white.

313 *Mucronella pendula*

Fruit bodies up to 30 mm long, pendulous on a short stalk, translucent white becoming yellowish with age. Colonies of these delicate 'stalactites' are found on the underside of or in hollows of rotting logs in wet forest and rainforest. Uncommon. **Spores** c. 6.5 x 5 µm, ovoid to subglobose, smooth. Spore print white.

312

313

314 *Pterula.* aff. *stipata*

Fruit bodies up to 30 x 30 mm, the branches curved, attenuated, with pointed tips. Uncommon, found on bark of decaying logs in rainforest. Identification of the specimen photographed has not been confirmed microscopically, but the growth habit matches *P. stipata*. **Spores** not examined.

316

315 *Ramariopsis crocea*

Fruit bodies up to 50 mm high and 20–40 mm broad, with branch tips bluntly pointed or rounded. This bright little coral fungus is common and widespread in a variety of forest and heathland habitats in Australia and temperate regions of the world. **Spores** c. 4 x 5 µm, subglobose, ornamented with short spines.

316 *Ramariopsis* sp.

Fruit bodies up to 60 mm high, at first white, becoming pale yellow-brown. Found in open areas on the ground amongst leaf litter in tropical rainforest. **Spore** print white.

1.4.2 *Ramaria* and allies

317 *Beenakia dacostae*

Caps to 25 mm across, occasionally joined together, the texture soft with soft spines below. An uncommon but distinctive fungus found on dry debris under fallen eucalypt logs in wet forest. Its microscopic characters indicate a relationship to the genus *Ramaria*. **Spores** c. 9 x 3.5 µm, ellipsoidal or pip-shaped, finely warty. Spore print pale yellow-brown.

318 *Gomphus floccosus*

Fruit bodies to 150 mm high, fleshy, cylindrical at first, then expanding above to become deeply funnel-shaped. Fertile surface of broad, low, decurrent ridges. A rare fungus associated with plantations of introduced conifers. **Spores** c. 18 x 8 µm, elongate-ellipsoidal with coarse end ridges. Spore print yellow-brown.

317

318

319 *Gomphus* sp.

Fruit bodies up to 200 mm across, fleshy, with irregular, overlapping, spreading lobes arising from a common base. The fertile undersurface (**319b**) has thick, obtuse, decurrent ridges with numerous cross ridges. A few species of *Gomphus* occur in association with *Nothofagus* (myrtle beech) in southern Australia, especially Tasmania, and are yet to be described. **Spores** c. 25 x 10 µm, long-ellipsoidal with large irregular ridges. Spore print yellow-brown.

320 *Ramaria botrytis var. holorubella*

Fruit bodies up to 100 mm high, the branches are crowded and have rounded pink to wine red tips. A robust, short-stemmed coral fungus, fairly common in eucalypt and mixed forest. **Spores** c. 11 x 4 µm, ellipsoidal, curved, finely warty. Spore print clay-coloured.

321 *Ramaria capitata*

Fruit bodies up to 100 mm or more high, densely branched with rounded, viscid tips compressed together to produce a coarse, cauliflower-like texture. Colour variable from cream to pale yellow or pinkish orange. Widespread in forest and heathland. **Spores** c. 12 x 4.5 µm, ellipsoidal, finely warty. Spore print pale yellow-brown.

322 *Ramaria sp*

Fruit bodies up to 30 mm high, at first white, becoming cream-buff, the tips sometimes divided. Found amongst forest litter, particularly decaying tree-fern debris. **Spores** c. 6 x 3.5 μm, ellipsoidal, finely warty. Spore print yellow-brown.

323 *Ramaria flaccida*

Fruit bodies up to 55 mm high, usually in colonies, often in 'fairy rings', or arcs. Found in eucalypt forest and teatree thickets, and sometimes in pine plantations. **Spores** c. 8 x 4 μm, ellipsoidal, finely warty. Spore print pale yellow-brown.

324

325

326

327

324 *Ramaria* aff. *formosa*

Fruit bodies to 150 mm high, sometimes massive and forming rings or colonies. Widespread in *Nothofagus* and *Eucalyptus* wet forest. Spores not examined.

325 *Ramaria gracilis* [= ***Ramaria filicicola***]

Fruit bodies up to 100 mm high, usually densely branched, the branches slender, white to clay-coloured, becoming darker with age. A common fungus of eucalypt and pine forest, where it frequently forms large expanding growth rings or arcs several metres in diameter. **Spores** c. 6 x 4 µm, ovoid, minutely rough. Spore print yellow-brown.

326 ***Ramaria lorithamnus*** [= *Ramaria sinapicolor*]

Fruit bodies up to 100 mm high, usually forming densely caespitose clumps. Individual fruit bodies with one to three branches, bruising to pinkish brown. Found in tall eucalypt forest and rainforest. **Spores** c. 8.5 x 5 µm, ellipsoidal with low irregular warts. Spore print yellow-brown.

327 *Ramaria anziana*

Fruit bodies up to c. 150 mm high, the tips of branches acute or rounded, sometimes compressed and cauliflower-like in texture, yellow. Stems and branches salmon-pink or orange. It frequently forms large rings or arcs in eucalypt forest as the feeding mycelium radiates outwards. **Spores** c. 10 x 5 µm, ellipsoidal, finely rough. Spore print yellow-brown.

328 *Ramaria versatilis* (previously included in *R. fumigata*)

Fruit bodies up to 100 mm high, densely branched and varying in colour from lilac to greyish purple, often with brownish tinges from maturing spores. Widespread in eucalypt and mixed forest. POISONOUS. **Spores** c. 10 x 4.5 µm, ellipsoidal with pointed ends, finely rough. Spore print yellow-brown.

1.5 PUFFBALLS AND ALLIES

1.5.1 Simple Puffballs

The members of this group are the common puffballs of grassland and forest. Some species grow singly, others occur in dense clusters. When young most are covered with a layer of soft warts or spines that fall away as the puffball matures. Spores are discharged through an apical pore when pressure is applied to the outer membrane.

329 *Lycoperdon perlatum*

Fruit bodies to 60 mm across. The soft outer spines become darker and fall off as the puffballs expand. Mature to old specimens darken to olive brown and become shiny. Usually found in small clusters amongst leaf litter in forest and woodland. **Spores** c. 4 µm diam., globose, finely roughened. Spore mass brown.

329

330 *Lycoperdon pyriforme*

Fruit bodies to 40 mm across, club or pear-shaped, almost white at first and covered with soft, pale brown deciduous warts. Mature puffballs are smooth and brown with an apical pore. Grows in clumps on rotting wood. **Spores** c. 4 µm diam., globose, smooth. Spore mass olive brown.

331 *Lycoperdon scabrum*

Fruit bodies to 40 mm. across, at first densely covered with soft dark fibrillose pointed scales that fall off as the 'ball' expands, leaving a smooth, shiny surface skin. Occurs on the ground in open forest and woodland. **Spores** c. 5 µm diam., globose with long stem, rough. Spore mass brown.

332 *Vascellum pratense* [= ***Lycoperdon pratense***]

Fruit bodies to 50 mm across, when young covered with fine white warts that fall off with growth, exposing the white skin that darkens with age. Spores are discharged through a large torn orifice (**332b**). Found in rings amongst grass. **Spores** c. 4 µm diam., globose. Spore mass pale brown.

1.5.2 Earth Stars

Members of this group have a double layer of tissue, the outer layer of which splits to expose the central puffball containing the spores.

333 *Geastrum fornicatum* Arching Earth Star

Fruit bodies to 60 mm across. At maturity the outer lobes curl under to elevate the spore sac. A distinctive fungus usually found in arid areas. **Spores** c. 5–7 µm diam., globose. Spore mass brown.

334 *Geastrum triplex* Earth Star

Expanded fungus variable in size from 70 mm to 100 mm across the tips of the lobes. The young specimens appear as small half-buried puffballs; at maturity the outer layer splits in the familiar stellate manner to reveal an inner, thin-skinned sac containing the spores. Widespread in most forest habitats. **Spores** c. 5 µm diam., globose, spiny. Spore mass dark brown.

1.5.3 Hard-skinned Puffballs

Mycenastrum, *Nothocastoreum* and *Scleroderma* have a tough outer layer that splits into lobes at maturity to expose the spores for dispersal by wind and rain. *Pisolithus* erodes from the top as the spore cells mature.

335 *Mycenastrum corium*

Fruit bodies to 200 mm across, globose to subglobose or pear-shaped. This is a large puffball of desert and semi-desert regions and occasionally of coastal dunes. When mature, the tough outer layer splits into lobes exposing the dark, spore-containing gleba. **Spores** c. 12 µm diam., globose and densely spinose. Spore mass dark brown.

335

336 *Nothocastoreum cretaceum*

Fruit bodies to 15 mm across, subglobose, at first almost buried, becoming exposed at maturity when it splits into several lobes to release the powdery spore mass. The outer shell is hard and brittle. Usually in half-buried clusters in eucalypt forest. Commonly seen after fire. **Spores** variable, c. 14 x 6 µm, ellipsoidal with distinct stalk, coarsely roughened, colourless.

337 *Pisolithus arhizus* Horse Dung Fungus

Fruit bodies variable, to 120 mm across, may grow singly but usually in clumps. The skin of young fruit bodies is brown and shiny with an irregular mottled dark pattern. Spores develop in the cells shown in the cut specimen and mature from the top downwards, the skin progressively eroding. It is usually seen along country road edges. **Spores** variable, to 12 µm diam., globose, densely covered with spines. Spore mass brown.

338 *Pisolithus* sp. [= ***Pisolithus albus***]

Fruit bodies variable, with a growth habit similar to *Pisolithus arhizus*, but the outer skin is white, with no mottling pattern, even at maturity, and erodes in larger flakes. The spore-containing cells are also smaller. It appears to be the common form in mallee and desert regions. **Spores** variable, up to 12 µm diam., globose, densely covered with spines. Spore mass brown.

339 *Scleroderma cepa*

Fruit bodies to 60 mm across, solitary to gregarious, sometimes clustered. The smooth surface of young specimens becomes reticulated or cracked with age. The outer skin splits into recurved lobes when mature, exposing the spore mass. Common in mixed forest. **Spores** c. 24 µm diam., globose, densely spiny. Spore mass dark brownish purple.

340 *Scleroderma paradoxum*

Fruit bodies c. 15 mm across, subglobose, sometimes lobed, hypogeous, occurring singly or in clusters. This fungus is rarely seen, but is occasionally found unearthed by lyrebirds and fungivorous animals in tall eucalypt forest and rainforest. **Spores** variable, 8–20 µm diam., globose, covered with broad based or slender dark spines.

339

340

1.5.4 Bird's-nest Fungi

Most members of this group produce their spores in peridioles contained in open nest-like structures. The peridioles are usually scattered by the action of water drops falling into the 'nest'.

341 *Crucibulum laeve* and *Nidula emodensis*

Two species of fungi are illustrated growing together. *Crucibulum laeve* has larger fruit bodies, up to 12 mm across, flared, with peridioles c. 2 mm across. *Nidula emodensis* has fruit bodies to 6 mm across that are externally hairy with darker peridioles c. 0.5 mm across. Both species grow on decaying wood and organic matter. The peridioles are splashed out by falling water drops.
Spores for both species c. 8 x 5 μm, ellipsoidal, smooth, colourless.

342 *Cyathus stercoreus*

Fruit bodies to 12 mm across with flared rims at maturity. Young specimens densely covered with reddish brown hairs. The fruit bodies often grow in densely crowded colonies with individuals distorted by mutual pressure. The inner surface of the 'cups' is a polished lead-grey. The lenticular peridioles are shining black, c. 2 mm diam. Found on herbivore dung. **Spores** variable, 20–40 µm, globose, smooth, colourless.

343 *Sphaerobolus stellatus* Cannon-ball Fungus

The tiny star-shaped fruit bodies, c. 1.5 mm across, are partly immersed in a layer of the associated fungal tissue. At maturity, the brown peridioles, c. 1 mm. diam., are shot out by sudden inversion of the membrane lining the 'star' — a remarkable spore dispersal mechanism. Commonly found on decaying wood, bark and herbivore droppings. **Spores** c. 8 x 6 µm, broadly ellipsoidal, colourless.

1.5.5 Stalked Puffballs

344 *Battarrea stevenii* Mallee Drumsticks

Fruit bodies to 400 mm high, rising from a basal cup-like structure. The fertile head is covered by a cap that falls off at maturity exposing the powdery spore mass. The stem is hard and woody. Found in arid regions associated with mallee eucalypts and melaleucas and in coastal melaleuca thickets. **Spores** c. 6 µm diam., globose to subglobose with fine spots. Spore mass rust-brown.

344

347

345 *Calostoma fuhreri*

Fruit bodies up to 15 mm high when dry, expanding to c. 35 mm when hydrated. Fruit bodies occur in clusters that recede into their own earth cavities when dry, emerging when wet. The longer spores and arid habitat separate it from the similar species *C. rodwayi*. The name *Calostoma*, meaning beautiful mouth, refers to the coloured orifice when newly exposed. **Spores** c. 10 x 28 µm, cylindrical, reticulated. Spore mass white.

346 *Calostoma fuscum*

Fruit bodies to 60 mm high. The stem supports a three-layered peridium. The outer layer is a protective cap that falls off at maturity exposing a thin-walled puffball with an orange-margined stoma. The spore sac, suspended below the stoma, releases spores through its opening when pressure is applied to the puffball surface. Widespread in eucalypt forest. **Spores** c. 12 x 9 µm, ellipsoidal or cylindrical with rounded ends, reticulated. Spore mass white.

347 *Calostoma rodwayi*

Fruit bodies to 25 mm high. The head, c. 10 mm diam., bears the stoma that is covered by a lobed cap that falls off at maturity. The spore sac, suspended beneath the stoma, releases spores when pressure is applied to the puffball. Found on the ground and on clay banks in fern gullies and rainforest. **Spores** variable, c. 10–18 x 8–15 µm, ellipsoidal to subglobose, reticulated, white.

348

349

348 *Calostoma* sp.

Fruit bodies to 60 mm high; they resemble *Calostoma fuscum*, but are paler and somewhat thicker and the outer cap is more gelatinous, exposing a translucent puffball through which the spore sac is visible. The stoma is a raised cross, and pale compared with other species. It is known from the Victorian Grampians and coastal heathland to the south. **Spores** c. 10–15 x 8–10 µm, ellipsoidal to subglobose, reticulated, white.

351

352 →

349 *Phellorinia herculeana* subsp. *herculeana*

Fruit bodies to 120 mm high, fertile head ellipsoidal, up to 45 mm diam. and expanding from a thick stem, white when fresh and covered with large overlapping scales. Found in mallee and desert habitats. **Spores** c. 7 µm diam., globose, covered with small spines. Spore mass yellow-brown.

350 *Phellorinia herculeana* subsp. *strobilina*

Fruit body to 100 mm high, the club-head is more expanded than is subsp. *herculeana*, and has large pyramidal warts rather than overlapping scales. Found in mallee and desert habitats. **Spores** c. 7 µm diam., globose, covered with small spines. Spore mass yellow-brown.

351 *Podaxis beringamensis*

Fruit bodies to 350 mm high, at first covered with coarse scales that flake off at maturity to expose the spore mass beneath. It is much taller than its relative, *P. pistillaris*, and is found exclusively on termite mounds in northern arid regions of Australia. The fruit bodies illustrated are young specimens, not yet fully expanded. **Spores** c. 10 x 7.5 µm, ellipsoidal with flattened end, smooth. Spore mass dark brown.

352 *Podaxis pistillaris*

← Fruit bodies to 150 mm high. The gleba is covered by a cap that splits at maturity, falling away to expose the dark powdery spore mass. Found in deserts and semi-deserts of Australia and other countries. It is often abundant after rain. **Spores** c. 15 x 10 μm, ellipsoidal to ovate with stalk, smooth. Spor mass dark brown.

353 *Tulostoma* sp.

Fruit bodies 5–20 mm high depending on the species; peridium globose to subglobose on a slender stem that varies from buried to fully exposed. Although the genus is easily recognised, species are not easy to separate in the field. They are most common in sand and sandy soils in dry and arid regions. **Spores** not examined.

1.5.6 Stinkhorns

The Stinkhorns are a distinctive group having bizarre forms accompanied b strong, unpleasant odours. The fruiting bodies develop in egg-like sacs tha are ruptured by the spore-bearing receptacle as it rapidly expands at maturity The spore-bearing gleba is a foul-smelling brown slime that is eagerly con sumed by blowflies and other insects. Spores are distributed after passin through the insects.

354 *Aseroë rubra* Starfish Fungus

Fruit bodies to 130 mm across, with up to 8 arms, attached to a central disc that is covered by gleba. Usually found in alpine and subalpine grassland and open forest. **354a** illustrates a receptacle expanding from its encasing sac; **354b** shows a fully expanded specimen with central gleba. **Spores** c. 5 x 2 μm, ellip soidal, smooth.

354a

354b

355a

355b

355c

355 *Clathrus archeri* [= *Anthurus archeri*]

Fruit bodies to 160 mm measured across the 4 to 8 expanded arms that radiate from a short hollow basal tube. It is most common in alpine meadows and woodland but also grows on wood mulch, where fruit bodies grow larger, with more arms, some of which may have divided tips (**355b**). The malodorous gleba is borne on the upper surfaces of the arms. Flies feeding on the spore-bearing gleba are shown in **355c**. **Spores** c. 7 x 2.5 µm, ellipsoidal, smooth.

356 *Colus hirudinosus*

Fruit bodies to 70 mm high, variously joined at the apex. The gleba is borne on the inner surface of the 5 or 6 arms that join at the base to form a short tube. The species prefers moist or winter-wet depressions in open forest and woodland. Uncommon. **Spores** c. 6 x 2 µm, ellipsoidal, smooth.

357

359

357 *Ileodictyon cibarium* Lattice or Basket Fungus

At maturity the stemless receptacle of this fungus emerges from its 'egg', often quite suddenly. The gleba is borne on the inner surface of the lattice. Widely distributed in forests and in wood mulch on gardens, but not common. **Spores** c. 5 x 2 µm, ellipsoidal, smooth.

358

360

358 *Lysurus gardneri*

Unexpanded 'eggs' to 30 mm diameter. Spore-bearing receptacle to 150 mm high. The inner surfaces of the five to seven short arms bear the slimy, particularly foetid-smelling gleba. Found on organic material including rotting wood and herbivore dung. **Spores** c. 4.5 x 1.5 µm, ellipsoidal, smooth.

359 *Mutinus boninensis*

Fruit bodies to 60 mm high and 5 mm diam. arising from small egg-like sacs on loose net-like rhizomorphs on dead wood in wet forest. The fertile receptacle emerges rapidly from the 'egg' at maturity. The gleba forms a band around the apex of the receptacle. Found on decaying wood in wet forest. Uncommon. **Spores** c. 4.5 x 2.5 µm, ellipsoidal, smooth.

360 *Mutinus borneensis*

Unexpanded 'eggs' to 20 x 15 mm; expanded receptacles to 80 mm high with the gleba forming the band seen near the apex. The odour is less obnoxious than most members of the group. The specimen illustrated was growing in wood mulch on a garden bed in the Royal Botanic Gardens, Melbourne. Uncommon. **Spores** c. 3.5 x 2 µm, ellipsoidal, smooth.

361 *Mutinus cartilagineus*

Fruit bodies to 60 mm high with a firm rubbery texture. At maturity the expanding receptacle ruptures the outer skin exposing the olivaceous gleba. Occasionally the outer skin fails to rupture properly, producing curious sheathed forms. Solitary to gregarious on soil and humus. Uncommon. **Spores** c. 4 x 2 µm, ellipsoidal, smooth.

362 *Phallus multicolor*

Fruit bodies to 180 mm high, the stems to 3 cm diam. The indusium or 'crinoline' varies from white to shades of yellow or orange and can vary from short to full length. The gleba is produced in the depressions of the apex.
A tropical species found on organic debris in rainforest and on garden wood mulch and compost. **Spores** c. 4 x 2 µm, ellipsoidal, smooth.

363 *Phallus rubicundus*

Unexpanded 'eggs' to 25 mm diam.; receptacle to 150 mm high, solitary or in clusters. Colour can vary from orange to red. Usually found on organic humus and decaying wood debris in tropical and subtropical habitats.
Spores c. 4.5 x 2 µm, ellipsoidal, smooth.

362

Photo Ian McCann

364 *Pseudocolus fusiformis*

Fruit bodies to 60 mm high, with up to four arms, united at the apex. The inner surface of the arms bears the malodorous gleba. Found in tropical and subtropical rainforest on the ground and decaying forest debris. **Spores** c. 4 x 2 µm, ellipsoidal, smooth.

1.6 UNDERGROUND OR TRUFFLE-LIKE FUNGI

365 *Mesophellia glauca*

Fruit bodies to 30 mm diam. occur in clusters below the soil surface and are associated with the roots of the trees under which they grow. The thick, outer layer is brittle and coated with cemented sand grains. Beneath this is the spore-containing gleba. The inner core is firm and corky and is eaten by various small animals that are attracted by the strong garlic-like odour of mature specimens. Found in eucalypt forest, usually after fire. **Spores** c. 8 x 4.5 µm, ellipsoidal, smooth. Spore mass pale green-grey.

365

366 →

367

368

369

366 *Octaviania tasmanica*

Fruit bodies to c. 20 mm across, subglobose to somewhat flattened. Gleba pale at first, becoming dark and mottled with age. Found in small clusters just below ground in various forest habitats. **Spores** c. 12 µm diam., globose, densely and coarsely spiny. Spore print brown.

367 *Protoglossum luteum*

Fruit bodies to 15 mm across, subglobose to irregular with firm, gelatinous, transparent outer skin. Sections show irregular chambers lined with fertile tissue. Found in wet forest just below ground level. **Spores** c. 13 x 10 µm, ellipsoidal with stem, thick-walled, smooth. Spore print tan.

368 *Richoniella pumila*

Fruit bodies to 25 mm across, irregularly subglobose. Flesh fragile; sections show small chambers. Occasional in small clusters at or just below surface in various forest habitats. **Spores** c. 10 x 9 µm, almost cuboid, smooth. Spore print brownish pink.

369 *Rhizopogon luteolus*

Fruit bodies to c. 50 mm across, usually half-immersed in soil. Texture rubbery and sponge-like, internally consisting of small chambers. Immature specimens white inside but soft and brown when mature. Associated with pine trees and plantations, where often common. **Spores** c. 8 x 3 µm, ellipsoidal, smooth. Spore mass yellow-brown.

370 *Thaxterogaster* sp.

Fruit bodies to 20 mm across, unevenly subglobose. Sections show internal structure of plates and chambers. Found in colonies at or below the ground surface in rainforest habitats. **Spores** c. 11 x 6.5 µm, ellipsoidal to almond-shaped, warty. Spore print rust-brown.

371a

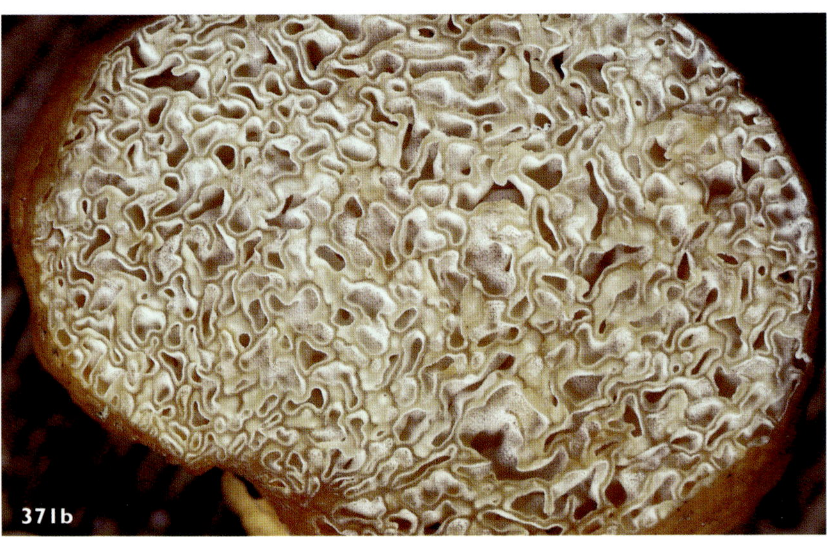
371b

371 *Zelleromyces* sp.

Fruit bodies to c. 35 mm across, irregularly subglobose, hypogeous or emergent at maturity. Inner tissue consists of irregular chambers c. 2 or 3 per mm (**371b**). A milky latex is produced when cut. Fruit bodies are frequently exposed by birds and other animals scratching for food. Found in various forest types. **Spores** c. 9 x 8 µm, subglobose, ornamented with amyloid ridges, white.

1.7 SPINE FUNGI

The fertile tissue in this group covers the surface of spines or teeth instead of gills or pores. Some are mushroom-shaped, either fleshy or tough and leathery, while others form closely adhering crusts on the undersides of logs and branches.

372 *Hericium coralloides*

Fruit bodies very variable, from a few branches with pendent spines to massive specimens up to 500 mm across and deep. It is a remarkably beautiful fungus with its multitude of spines managing to keep free from one another. Most commonly associated with *Nothofagus*, occasionally with other genera including *Acacia*, and introduced trees including oaks. **Spores** c. 4 x 3 µm, ellipsoidal with one side flattened, smooth. Spore print white.

372

373 *Hydnellum auratile*

Fruit bodies to 50 mm across, simple or with outgrowths, radially fibrillose-striate, concentrically zoned and sometimes fused with adjacent caps. It resembles a colour form of *Phellodon niger*, but can be easily separated by the spores. Found amongst leaf litter on the ground in tall eucalypt forest. **Spores** c. 5.5 x 4 µm, irregular with projecting flat-topped tubercles. Spore print brown.

374 *Hydnum repandum*

Caps to 100 mm across, convex, becoming irregularly contorted, often centrally depressed, sometimes with several caps fused together, the stems also occasionally fused at base. Flesh firm and brittle; fertile surface with brittle fleshy spines. Frequent on ground in forests. **Spores** c. 8 x 6 µm, ellipsoidal to subglobose, smooth. Spore print white.

373

374

375

376

375 Mycoacia subceracea [= *Phlebia subceracea*]

This species forms bright yellow patches on the undersides of fallen branches, sometimes extending for their entire length. The thin tissue is densely covered with short, soft spines. **Spores** c. 6 x 2.5 µm, ellipsoidal, smooth. Spore print white.

376 *Phellodon niger*

Fruit bodies to 50 mm across, simple or confluent, flat to funnel-shaped, radially silky-striate. Texture tough and leathery. Teeth usually pale grey-lilac when fresh. Dried specimens have a distinctive fenugreek-like odour. On the ground in forest and heathland. **Spores** c. 4 x 3 µm, broadly ellipsoidal, spiny. Spore print white.

377 *Sarcodon fuligineoviolaceus*

Fruit bodies to 60 mm across, occurring singly or fused together. The pinkish to lilac flesh stains blue-green on contact with a strong alkaline solution. Fresh spines are pale greyish lilac, becoming brown with age. Found in rainforest associated with *Nothofagus cunninghamii*. Uncommon. **Spores** c. 6 x 4 µm, irregular, roughly rounded, tuberculate. Spore print white.

378 *Steccherinum* sp.

Fruit bodies forming irregular colonies of varying size on the underside of decaying wood. The fertile tissue covers the soft spines. Identity uncertain. **Spores** not examined.

1.8 WOODY PORE-FUNGI & BRACKET-FUNGI (*Polyporus* and allies)

This group of fungi includes the leathery, tough or woody brackets, as well as mushrooms and adherent crusts that have pores or woody gill-like plates instead of the soft gill tissue of agarics and their relatives in Group 1.1. Some form mycorrhizal associations with trees, while others consume dead wood and structural timbers. Other species are important tree pathogens, causing disease and death of the host.

379 →

380

381a

381b

379 *Amauroderma rude*

← Caps or brackets to 150 mm across, somewhat polished in appearance, with radial and concentric ridging. Texture hard and woody. Stems central or lateral, dark brown, velvety. The white, pored undersurface rapidly turns dark reddish brown when bruised. Expanding fruit bodies frequently envelop grass and twigs. Found near decaying stumps and buried wood. **Spores** c. 9 x 7 µm, ovoid, finely dotted. Spore print brownish yellow.

380 *Antrodiella citrea*

Fruit bodies soft and chamois-like in texture, adhering to the underside of dead twigs and branches. It forms soft narrow shelves up to c. 20 mm wide and of varying length. The upper surface is bright yellow, the pore surface white. Common on dead eucalypt twigs and branches. **Spores** c. 4.5 x 2.5 µm, ellipsoidal, smooth. Spore print white.

381 *Antrodiella zonata* [= *Irpex brevis*]

Fruit bodies form extensive shelving brackets arising from a closely adhering base. Fertile tissue with irregular tubular or flattened tooth-like pores. Large bare or crystal-coated metuloids are abundant on the fertile surface (**381b**). It colonises a range of hardwood and softwood tree remains. **Spores** c. 4 x 2 µm, ellipsoidal, smooth. Spore print white.

382 *Aurantiporus pulcherrimus* Strawberry Bracket Fungus

Brackets to 120 mm wide and 80 mm radius, soft and spongy. It forms single or compound brackets from a broad base, or occasionally longer fused shelves on dead logs or living trees. It sometimes forms a closely adherent layer on the underside of logs. It usually occurs on *Nothofagus cunninghamii* (myrtle beech) and occasionally in alpine country on snow gums. **Spores** c. 5 x 4 µm, subglobose, smooth. Spore print white.

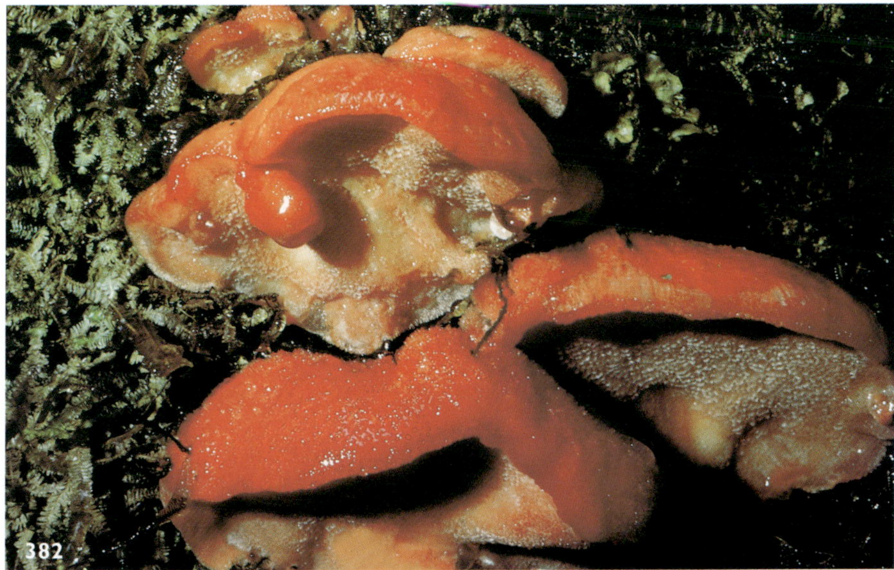
382

383 *Australohydnum dregeanum* [= *Irpex griseofuscescens*]

Fruit body forms adhering layers of purplish-grey tooth-like projections from a broad base on surfaces and undersides of dead tree trunks and branches. The teeth have an abundance of crystal-covered metuloids on their surface. **Spores** c. 4.5 x 2 µm, smooth. Spore print white.

384 *Australoporus tasmanicus*

Fruit bodies form irregular, often shelving layers and brackets on living and dead tree trunks. It causes a heart-rot when growing on living trees. A common host is *Nothofagus cunninghamii* (myrtle beech). **Spores** c. 8 x 4.5 µm, ellipsoidal, smooth. Spore print white.

383

385 *Coltricia cinnamomea*

Fruit bodies to 30 mm across, single or in small colonies. The distinctive satin-like lustre is caused by the pattern of shining radial surface hairs. The pore surface is rusty brown, pores 1–3 per mm. It grows from decaying sub-surface wood and tree roots in forest and woodland. **Spores** c. 7 x 4.5 µm, broadly ellipsoidal, smooth. Spore print yellow-brown.

386 *Coltriciella dependens*

Fruit bodies variable, from 5–25 mm across, irregular, pendent on short stems from the underside of decaying or burnt logs and branches in eucalypt forest. Sometimes the fruit bodies are attached without a distinct stem. Occasionally several caps are fused together. Pores angular, rust-brown, 2 or 3 per mm. **Spores** c. 8 x 5 µm, ellipsoidal to broadly pip-shaped, finely warty. Spore print yellow-brown.

387 *Fistulina hepatica* Beefsteak Fungus

Fruit bodies to 100 mm across, attached by a short lateral stem. Texture firm-fleshy. The fertile under surface consists of a layer of separate tubes that are pink when mature, a distinctive feature of this species. It is found on logs and stumps and occasionally at the base of living trees. **Spores** c. 4–5 µm diam., subglobose, smooth. Spore print pink.

388 *Fomes hemitephrus*

Fruit bodies to 200 mm or more across, with white pore surface beneath, on living or dead trees. The brackets have a rather acute edge. A distinctive orange layer occurs just below the upper surface. A cause of heart-rot in living trees, particularly *Nothofagus cunninghamii* (myrtle beech). Frequent in wet forest. **Spores** c. 5 x 2 µm, oblong-ellipsoidal, smooth. Spore print white.

387

388

389a

389b

389 *Fomitopsis lilacinogilva*

Fruit bodies to 160 mm wide and 80 mm radius from a broad lateral base, often forming overlapping shelves. Upper surface furrowed, from pale to dark shades of brown with a rounded pinkish margin. Fresh pore surface pale lilac, bruising to deep pinkish-lilac. Found on several species of dead wood including red gum sleepers, log barriers and structural timbers, causing a brown cubical rot (**389b**). **Spores** c. 7.5 x 2.5 µm, broadly ellipsoidal, smooth. Spore print white.

390 *Ganoderma applanatum* [= ***Ganoderma australe***]

Fruit bodies to 600 mm or more across. A new fertile layer is added each year. The new pore layer is white and readily bruises dark brown when scratched or bruised. Messages written on the white pore layer become permanent when the fungus dries. Occurs on living or dead trees in wet sclerophyll forest and rainforest. Causes a white heart-rot. **Spores** c. 10 x 6 µm, ovoid, finely spotted or roughened. Spore print brown.

391 *Ganoderma resinaceum* [= ***Ganoderma weberianum***]

Fruit bodies to 250 mm wide, 140 mm radius, the surface nearly flat, polished, with concentric pattern of reddish brown bands, the outer bands dull brown. Pores white, becoming brown with age. Found on tree trunks in tropical forest, apparently named for its resinous upper surface. **Spores** c. 12 x 9 µm, ellipsoidal, rough. Spore print brown.

390

391

392 *Gloeophyllum abietinum*

Fruit bodies to 80 mm or more wide, sometimes forming continuous shelves. It is found on dead conifer wood. It differs from its close relative *G. sepiarium* in having more widely spaced gills, longer spores and duller brown colours lacking yellow. Common in mallee habitats on *Callitris* spp. **Spores** c. 12 x 4 µm, cylindrical to slightly curved, smooth. Spore print white.

393 *Gloeophyllum sepiarium*

Fruit bodies to 70 mm across, radius to 50 mm, with a distinctive fertile surface. It grows mainly on dead conifer wood including native species such as Cyprus pine, and pine building timber in moist conditions. **Spores** c. 10 x 4 µm, cylindrical, smooth. Spore print white.

394 *Gloeoporus phlebophorus*

Fruit bodies to 35 mm wide, 25 mm radius, annual. Forms single brackets to imbricate colonies of many brackets arising from a common base. The brackets have a smooth waxy texture and, when dry, are hard and horny. Pores minute, up to 7 per mm. The most common host is *Nothofagus cunninghamii* (myrtle beech), but it is also recorded on other trees. **Spores** c. 3 x 1 µm, curved or sausage-shaped, smooth. Spore print white.

394

395

396

395 *Grifola colensoi*

Fruit bodies variable, large, compound, with several caps on thick, rubbery branches arising from a common lateral to central base. Sometimes forming dense raised rosettes. Pores decurrent, irregular and shallow. Found at the bases of living trees, usually *Nothofagus* and *Eucalyptus*. **Spores** c. 5 µm diam., globose, smooth. Spore print white.

396 *Haddowia longipes*

Caps to 120 mm across, with concentric zonation and shiny appearance. Stem to 160 mm long. Pore surface white to ochraceous; pores angular, 2 or 3 per mm. The distinctive spores are ornamented with broad longitudinal crests. Usually found in tropical forest and rainforest. **Spores** c. 16 x 12 µm, broadly ellipsoidal with longitudinal crests with cross-walls. Spore print brown.

397 *Hapalopilus nidulans*

Fruit bodies to 100 mm wide, roughly semicircular with broadly attached base. Pore surface cinnamon-coloured; pores angular, c. 3 or 4 per mm. A useful diagnostic character is the strong violet colour produced when a strong alkali solution is applied to the tissue. Found on dead logs and branches. **Spores** c. 4.5 x 2.5 µm, ellipsoidal. Spore print white.

398 *Hexagonia apiaria* [= ***Trametes apiaria***]

Fruit bodies to 100 mm wide and 80 mm radius, attached by a lateral base. Surface covered with purplish brown, branched hairs. The distinctive, large, angular pores are pale-edged when young. It is found on dead branches of a variety of tropical and subtropical trees. **Spores** c. 14 x 5 µm, cylindrical, smooth. Spore print white.

399b

399 *Hexagonia tenuis*

Brackets annual, to 100 mm wide, 80 mm radius and c. 3 mm thick, attached by a narrow lateral base. Upper surface often radially grooved. Pores shallow, angular, c. 8–10 per cm. Found on dead logs and branches of a variety of tree species. **Spores** c. 6 x 7 µm, narrow-ellipsoidal, smooth. Spore print white.

400 *Hexagonia vesparia* [= ***Osmoporus vesparius***]

Fruit bodies to 60 mm wide and 40 mm radius, hoof-shaped, attached by a broad lateral base. A fresh developing bracket is shown. The very large, thick-walled pores and pale upper surface darken with age. Found on dead and dying eucalypt trunks and branches. **Spores** c. 18 x 7 µm, cylindrical with rounded ends, smooth. Spore print white.

401 *Inonotus nothofagi*

→ Fruit bodies variable, often forming several overlapping shelves arising from an extensive adherent base. Brackets are often radially and concentrically grooved with a distinctive pale, crenulate margin. Pore surface at first greyish, becoming rusty brown. Confined to *Nothofagus* wood. **Spores** c. 5.5 x 3.5 µm, ellipsoidal, smooth. Spore print rust-brown.

400

401 ←

402 *Laccocephalum hartmannii*

Fruit bodies to 150 mm across with central to lateral stems producing round to kidney-shaped caps. Young caps reddish-brown with a finely matt or velvet surface, becoming paler and cracking with age or on drying. Fruit bodies arise from underground pseudosclerotia at the base of trees. Uncommon, but may fruit freely after fire, when abnormally developed fruit bodies, lacking developed pores, are sometimes formed. **Spores** c. 9 x 3.5 µm, elongate-ellipsoidal, smooth. Spore print white.

403 *Laccocephalum mylittae* Native Bread

Fruit bodies to 120 mm across, round to irregular, single, or up to three together and freely produced after fire. The fruit bodies arise from an underground sclerotium consisting of dense tissue with a consistency resembling compacted boiled rice with an outer dark brown skin (**403b**).
The sclerotium was used as food by Aborigines. Widespread in eucalypt forest. **Spores** c. 6 x 2.5 µm, narrow-ellipsoidal, smooth. Spore print white.

402

403a

403b

404 *Laccocephalum sclerotinium*

Fruit bodies to 40 mm across, but usually smaller, arising from a small hypogeous sclerotium. This is one of several sclerotia-forming polypores that fruit freely after bushfires. The small fruit bodies are conspicuous on the bare, blackened forest floor. **Spores** c. 5 x 3.5 µm, ellipsoidal, smooth. Spore print white.

405 *Laccocephalum tumulosum* Stonemaker Fungus

Caps up to 140 mm across, at first white becoming grey or brownish grey with age. Fruit bodies arise from hypogeous, stone-like pseudosclerotia (**405b**). These are sometimes very large, giving rise to several fruit bodies. It commonly occurs in tall forests where it fruits freely after fire. **Spores** c. 10 x 3 µm, narrow-ellipsoidal, smooth. Spore print white.

405a

406 *Laetiporus portentosus* White Punk

Fruit bodies to 350 mm wide and 250 mm radius. Annual. Upper surface pale biscuit brown, smooth. Pore surface yellow when fresh, becoming dingy white. The soft spongy flesh is the food of certain insect larvae, and old and fallen brackets are riddled with the resulting tunnels. A distinctive species favouring living *Eucalyptus* trees, where it causes white heart-rot. Dry, smouldering brackets were reportedly used by Aborigines to transport fire. **Spores** c. 8 μm diam., globose to subglobose, smooth. Spore print white.

407 *Lenzites vespacea*

Fruit bodies commonly c. 100 mm wide and 30 mm radius, but sometimes much larger; broadly attached. Named for the resemblance of its pore structure to the cells of a wasps' nest. It is a tropical species that colonizes dead standing or fallen wood in rainforest. **Spores** c. 7 x 3 µm, cylindrical, smooth. Spore print white.

408 *Microporus affinis*

Fruit bodies to 100 mm wide and c. 80 mm radius, flabellate or fan-like. Colours range through various shades of brown in a strongly concentric banded pattern; the growing edge is paler. Mainly tropical, on fallen or standing dead wood. **Spores** c. 4 x 2 µm, ellipsoidal, smooth. Spore print white.

407

408

409 *Microporus xanthopus*

Funnels to 90 mm across, their inner surface distinctly zoned. Pore surface white to pale buff with minute pores, c. 10 per mm. The specific name *xanthopus*, meaning 'yellow foot', refers to the yellow-coloured disc of attachment. Found on dead logs and branches in tropical Australia.
Spores c. 2 x 4.5 µm, broadly ellipsoidal, smooth. Spore print white.

411

412

410 *Neolentiporus maculitissimus*

Brackets to 200 mm wide and c. 150 mm radius with single or multiple brackets adhering by a broad base to tree trunks and logs. Upper surface ornamented with numerous brown scales in concentric patterns. Pores small. Found in wet forest and rainforest. Causes brown heart-rot. **Spores** c. 15 x 4 µm, cylindrical, smooth. Spore print white.

411 *Phaeotrametes decipiens*

Fruit bodies to 50 mm wide, broadly adhering to the underside of host branches, or forming hoof-shaped brackets on the sides of host wood. The brown colour is sometimes tinged with violet. Pores c. 1 or 2 per mm. Usually found on species of she-oak in semi-arid to arid habitats. **Spores** c. 16 x 6 µm, ellipsoidal, smooth. Spore print yellowish white.

412 *Phellinus wahlbergii*

Brackets to 250 mm wide, 100 mm radius, producing single or multiple shelves broadly attached to the host wood. This fungus decomposes stumps and logs, and produces a white pocket rot. It also causes heart-rot in living trees. Pores small, 6–8 per mm. Usually in wet forest and rainforest. **Spores** c. 5 x 3 µm, ovoid to subglobose, smooth. Spore print yellow-brown.

413 *Piptoporus australiensis* Curry Punk

Fruit bodies to 200 mm wide and 120 mm radius, or several joining to form long shelves on fallen tree trunks and logs. The common name refers to its persistent curry-like odour, even when old and dry. It has a strong orange staining pigment. Causes a brown cubic rot. Widespread in eucalypt forest. **Spores** c. 9 x 6 µm, ellipsoidal, smooth. Spore print white.

414 *Polyporus arcularius*

Caps to 50 mm across, convex to centrally depressed or funnel-shaped on a central stem. It is characterized by the medium-sized coffin-shaped pores. The thin caps are tough and leathery when fresh, becoming brittle when dry. Fruit bodies solitary or gregarious on dead wood in various forest habitats. **Spores** c. 8 x 2.5 µm, cylindrical, smooth. Spore print white.

415 *Polyporus badius*

Fruit bodies to 150 mm across, with one to several brackets arising from a common base. Surface at first pale brown, becoming darker with age. Pores minute, c. 6–8 per mm. Stem usually short, brown-black, felty. The fungus resembles *P. melanopus* but is larger, and the cap more irregular and undulate with smaller pores. Found on dead tree trunks. **Spores** c. 7.5 x 3.5 µm, cylindrical-ellipsoidal, smooth. Spore print white.

413

414

415

416

417

416 *Postia caesia*

Brackets c. 50 mm wide and 40 mm radius with broad lateral attachment or sometimes dorsally attached to the underside of dead wood. Surface at first velvety, becoming smooth. Texture soft, sponge-like when fresh, staining blue to blue-grey when bruised or on ageing. Firm when dry. Pore surface white to blue-grey, pores c. 4 per mm. Widespread in various forest types, including pine forests. **Spores** c. 4.5 x 1 µm, cylindrical-curved, smooth. Spore print white.

417 *Postia lactea*

Brackets to c. 50 mm wide and 40 mm radius, white with spongy texture when fresh, firm and brittle when dry. Upper surface densely hairy. Pore surface white or pale cream, pores 4 or 5 per mm. Occurs on logs and branches of a variety of native and introduced trees. Does not stain blue as in *Postia caesia*. **Spores** c. 5 x 1.5 µm, cylindrical-curved. Spore print white.

418 *Postia pelliculosa*

Brackets to 80 mm wide and 60 mm radius, attached by a broad base c. 20 mm thick. Soft and moist when fresh, brittle when dry. Characterised by the dense red-brown layer of coarse hair on the upper surface. Pore surface white, bruising brown. Pores angular, variable, 2–5 per mm. Found on dead logs and stumps of eucalypts and other trees. **Spores** c. 5 x 3.5 µm, broadly ellipsoidal, smooth. Spore print white.

419 *Pycnoporus coccineus*

Brackets to 100 x 50 mm but variable, sometimes fusing to form longer shelves. Occasionally several brackets are produced from a single broad base. A common wood-decaying fungus. The brightly coloured red brackets are a common sight on dead logs and branches from moist fern gullies to semi-desert. **Spores** c. 3.5 x 4.5 µm, ellipsoidal, smooth. Spore print white.

418

419

420 *Rigidoporus laetus*

Brackets to 100 mm wide, radius up to 70 mm, single or with few to several brackets arising from a broad base. This species is found on tree trunks, stumps and dead logs. Large fluid drops are frequently seen suspended from the developing pore surface. **Spores** c. 5 µm diam., globose to subglobose, smooth. Spore print white.

421 *Ryvardenia campyla*

Brackets to 150 mm or more across, solitary, or compound with several brackets arising from a common base. Very variable but recognisable by texture, colour and habit. The fresh pore surface is frequently ornamented with clear or amber drops of fluid. Found on logs and stumps, and occasionally trunks of living trees. **Spores** c. 6 x 3.5 µm, ellipsoidal to pip-shaped, smooth. Spore print cream-coloured.

421

422 *Ryvardenia cretacea*

Brackets to 200 mm wide and 120m radius, convex to flattened, solitary or occasionally fused. Flesh dense and moist, drying to a soft chalky texture. Found on living tree trunks and dead logs in wet forest and rainforest. In southern Australia it is most frequent on *Nothofagus cunninghamii* (myrtle beech) and mountain ash. **Spores** c. 6 x 4 µm, pip-shaped, smooth. Spore print white.

423 *Trametes hirsuta*

Brackets to 80 mm wide and 50 mm radius. A robust member of the genus found on dead trunks, stumps and fallen branches of a variety of hardwood and softwood trees. The densely hairy region at the base of the brackets is a useful field character. **Spores** c. 8 x 3 µm, ellipsoidal, smooth. Spore print white.

424

426

424 *Trametes versicolor* Rainbow fungus

Brackets to 70 mm wide and 50 mm radius, often larger and forming flexible shelves. An active wood-decaying fungus, often forming extensive tiers on dead logs, branches and structural timber such as fence posts. It can also parasitise living trees. The underside is cream in colour, with a layer of tiny pores. It is a common and cosmopolitan species with a variety of colour forms. **Spores** c. 5 x 2 µm, narrow-ellipsoidal, smooth. Spore print white.

425 *Trametes* sp.

Probably a colour form of *T. versicolor*, differing in showing less zonation, and being much darker, with an obvious velvety pile on the upper surface. The pore surface is white. Found in forest gullies. **Spores** not measured. Spore print white.

426 *Tyromyces merulinus* [= ***Ceriporiopsis merulinus***]

Fruit bodies usually form extensive adhering colonies or patches c. 3 mm thick on the undersides of fallen logs and branches of various trees and shrubs. Pores angular, c. 5 per mm. It appears to prefer dead wood of *Nothofagus cunninghamii* (myrtle beech). **Spores** c. 4 x 2 µm, cylindrical to slightly curved, smooth. Spore print white.

425

1.9 LEATHERY SHELF-FUNGI

The common members of this group are thin and leathery, forming closely adhering crusts on dead wood. The crusts may be simple, forming little more than thin patches of colour on wood or, as in most of the larger species, free, flabby shelves or brackets that are hairy and zoned above but smooth on the lower, fertile surface.

427 *Aleurodiscus limonisporus*

Fruit bodies form irregular discoid colonies that are sometimes confluent, forming larger patches on dead tree trunks and branches. Fertile surfaces often have a powdery appearance because of a superficial layer of the very large spores. Fluid droplets are often present on the fertile surfaces. **Spores** c. 20 x 15 µm, lemon-shaped, smooth. Spore print white.

428 *Aleurodiscus* sp.

Fruit bodies to c. 4 mm across with a distinctive folded pattern; upper surface hairy. This tiny fungus occurs in wet forest on fronds and stems of ferns, particularly on bracken. Uncommon. **Spores** c. 8 x 4 µm, long ellipsoidal with one side flattened, smooth. Spore print white.

428

429

Byssomerulius corium

Fruit bodies form small to extensive adhering sheets, usually with shelving edges, on the underside of dead branches and twigs. The fertile surface forms a reticulated or finely wrinkled pattern. Cosmopolitan. **Spores** c. 7 x 3 µm, ellipsoidal, smooth. Spore print white.

430

432

433

431

430 *Cymatoderma elegans*

Fruit bodies to 200 mm across, funnel-shaped with distinctive external texture. Fairly common on decaying logs in tropical and subtropical forest in eastern Australia. **Spores** c. 6.5 x 4 µm, ellipsoidal, smooth. Spore print white.

431 *Dichostereum rhodosporum*

Fruit bodies form patches on the underside of bark and dead wood lying on the ground. Fresh fruit bodies have watery drops scattered over the surface. **Spores** c. 7 µm diam., globose, ornamented with low warts. Spore print white.

432 *Gloeoporus taxicola*

Fruit bodies form small patches to large sheets loosely attached to fallen branches, logs and tree stumps. The fertile surface has a reticulated pattern. Widespread on dead wood. **Spores** c. 4.5 x 1 µm, curved-cylindrical, smooth. Spore print white.

433 *Hyphodontia australis*

Fruit bodies form thin, closely adhering felt-like patches or sheets on the underside of dead, fallen or sometimes standing, wood. Mature specimens display a pattern of cracking and are covered with small irregular pimples. Strong alkali solutions produce an instant violet colour change. **Spores** not examined.

434 *Merulius tremellosus* [= *Phlebia tremellosa*]

This fungus forms gelatinous, adhering layers and shelves on the underside of decaying logs and branches and on decaying stumps and pine cones. It is most common in pine plantations. Colour is variable from creamy yellow to orange. **Spores** c. 3.5 x 1 µm, curved-cylindrical, smooth. Spore print white.

435 *Phlebia* sp.

Phlebia species are gelatinous in texture and adhere closely to the dead wood on which they grow. The fertile undersides typically form convoluted radiating patterns of tissue, becoming more or less flat at the margins. The tropical species illustrated was growing on dead wood. **Spores** not examined.

436

436 *Podoscypha petalodes*

Fruit bodies to 40 mm across, funnel-shaped, usually forming caespitose, rosetted colonies around trees, or on the ground growing from buried wood. Texture leather-like. Common. **Spores** c. 6 x 3.5 µm, ovoid, smooth. Spore print white.

437 *Podoserpula pusio*

Fruit bodies up to 100 mm or more high. This very attractive species has a unique appearance. A soft stem, usually pink, supports tiers of soft chamois-like lobes with folded pink undersides. It forms colonies on rotting tree trunks and logs, and occasionally on fallen tree-fern trunks. **Spores** c. 3.5 µm diam., globose, smooth. Spore print white.

438 *Serpula lacrimans* Dry Rot

Fruit bodies form sheets and irregular adhering patches on dead wood including structural timbers. The fertile surface is brown with a reticulated pattern. It is a very destructive fungus causing rapid breakdown of building timbers in cool, poorly ventilated, moist situations. **438b** shows fruiting structure on a boot. **Spores** c. 12 x 6.5 µm, ellipsoidal, smooth. Spore print brown.

437

438a

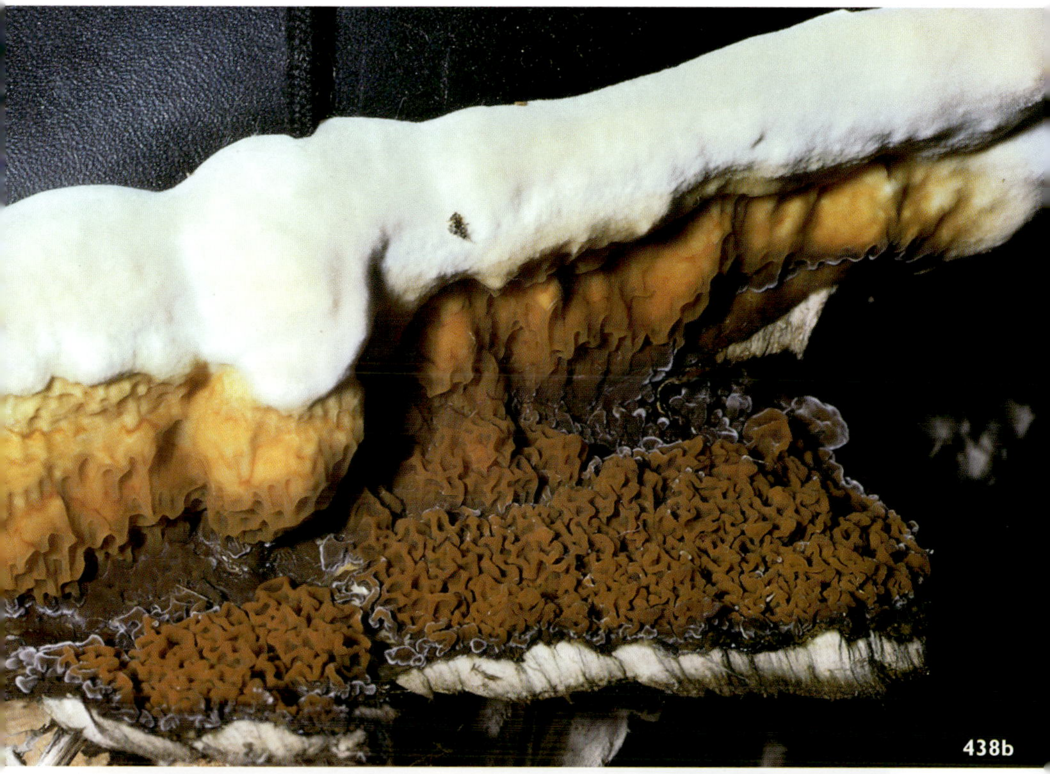

438b

439 *Stereum hirsutum*

This fungus forms extensive colonies on dead stumps and logs. The lower part of the thin leathery fruit body is closely adherent to the substrate. The shelving lobes are densely hairy on the upper surface. A common fungus on hardwood stumps and logs. **Spores** c. 6 x 3 µm, ellipsoidal, smooth. Spore print white.

440 *Stereum illudens*

A common and widespread fungus. It forms closely adhering, irregular violet patches on the underside of dead branches. The upper brown hairy surface is zoned (**440a**). The lower fertile surface (**440b**) is smooth, often with a pale margin. Frequent on dead logs and branches in eucalypt forest. **Spores** c. 8 x 3.5 µm, ellipsoidal, smooth. Spore print white.

441 *Stereum* sp.

These rhizomorphs are colonising fallen eucalypt bark.

442 *Stereum ostrea*

The fan-like lobes of this large, spectacular fungus colonise dead logs and branches, sometimes for much of their length. The fertile undersurface is a rich deep yellow-gold, the upper surface is distinctly zoned in brown and orange. A common species of wet to moist forest and moist habitats in dry forest. **Spores** c. 6 x 2 µm, narrow-ellipsoidal, smooth. Spore print white.

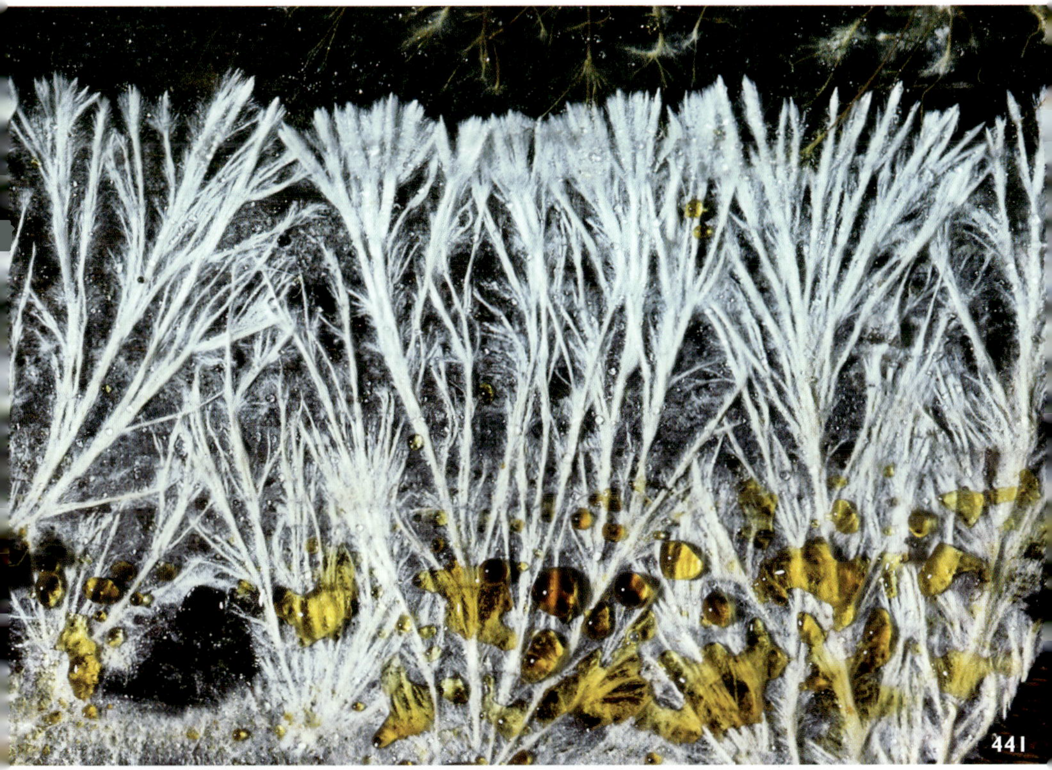

442

443

443 *Stereum rugosum*

This fungus is found on dead trees and decaying logs. The hairy, zoned upper surface is a rich chocolate brown. The fertile underside is grey, with paler margins. In the field it resembles *Xylobolus illudens*, which is purple on its underside. **Spores** c. 10 x 4 µm, curved, sausage-like, smooth. Spore print white.

444 *Thelephora terrestris*

Fruit bodies form irregular, often overlapping, rosettes and fan-like colonies on fallen pine needles and twigs in pine forest and under isolated pine trees. Easily recognised and common. **Spores** c. 9 x 7 µm, ovoid, finely angular and warty. Spore print white.

1.10 TRUMPET FUNGI

445 *Craterellus cornucopioides*

Fruit bodies trumpet-shaped, up to 40 mm across and 100 mm high. It usually grows with *Nothofagus* in Australian rainforest, but is widespread in the Northern Hemisphere under different tree types, especially beeches and oaks. Edible. **Spores** c. 10 x 15 µm, broadly ellipsoidal, smooth. Spore print pale cream-coloured.

445

446 *Craterellus* sp.

Fruit bodies to 50 mm across, funnel-shaped, but usually occurring as small clusters distorted by crowding. It could easily be mistaken for a large ascomycete (cup fungus) until examined microscopically. A tropical rainforest species. Uncommon. **Spores** c. 9 x 6.5 µm, broadly ellipsoidal to pip-shaped, smooth. Spore print white.

447 *Pseudocraterellus sinuosus* [= ***Pseudocraterellus undulatus***]

Fruit bodies trumpet-shaped, up to 30 mm across, sometimes larger, often with smaller outgrowths or proliferations. Usually growing in association with *Nothofagus* and other members of the family Fagaceae. Uncommon. **Spores** c. 10 x 7.5 µm, ovoid, smooth. Spore print white.

1.11 JELLY FUNGI

Members of this group, with the exception of *Septobasidium clelandii*, are gelatinous in texture and appearance. They have a very high water content, and are usually found on wood or as parasites of other fungi. The fertile tissue covers the greater part of the surface in convoluted forms, but only the lower surface of bracket forms. The basidia are quite different from other Basidiomycota. They are variously partitioned or forked.

448 *Auricularia auricula-judae* [= *Hirneola auricula-judae*]

Fruit bodies to 100 mm across, irregularly ear-shaped, gelatinous and translucent. Outer surface finely pubescent, the fertile surface smooth and shiny. Found on dead wood in tropical and subtropical forest. **Spores** c. 12 x 6 µm, cylindrical, curved, smooth. Spore print white.

448

449 *Auricularia polytricha* [= ***Auricularia cornea***]

Fruit bodies to 100 mm across, irregularly convex, contorted by crowding, densely pubescent above, the fertile surface smooth. Flesh thin, gelatinous, brownish red, translucent. Found on dead wood in tropical to subtropical forest, sometimes abundant on the eastern coast of Australia. **Spores** variable, up to 20 x 9 µm, cylindrical, curved, smooth. Spore print white.

450 *Calocera* sp. [= ***Calocera sinensis***]

Members of the genus *Calocera* are small, simple or branched jelly fungi. The colour ranges from yellow to orange or buff, according to the species. They are found on dead logs and twigs and in pine forest on fallen cones. **Spores** colourless.

449

450

451 *Heterotextus peziziformis* Golden Jelly-bells

Fruit bodies to 10 mm across, bell-shaped, with jelly-like texture. At first orange, becoming paler with age and drying orange-red. The shrunken dried fruit bodies rehydrate readily when moistened, quickly regaining their original size. A common fungus on dead logs and twigs in forests. **Spores** c. 15 x 5 µm, curved or sausage-shaped with several cross walls, smooth, colourless.

452 *Pseudohydnum gelatinosum*

Fruit bodies to 60 mm across, sometimes larger, or fused. The generic name *Pseudohydnum* refers to the resemblance of this fungus to species of *Hydnum* that bear spine-like projections on their fertile surfaces. A cosmopolitan species. Found on dead logs and branches and on the lower bark of living trees. **Spores** c. 6 x 5.5 µm, globose to subglobose, smooth, colourless.

453 *Septobasidium clelandii*

Fruit bodies to 15 mm long. This fungus is a remarkable insect parasite arising from a female gall-forming coccid bug, *Callococcus leptospermi*, that forms swellings on branches of *Leptospermum* species. The spores are produced on the black protruding structures. **Spores** c. 20 x 7 µm, cylindrical, curved, colourless.

452

453

454 *Sirobasidium brefeldianum* [= **Leucogloea compressa**]

Fruit bodies to 3 mm across, globose, white, translucent, forming small colonies to 3 mm across. Found on decaying wood in moist to wet shaded forest. **Spores** colourless, not measured.

455 ***Tremella encephala***

Fruit bodies variable, c. 15–25 mm across, irregularly rounded and folded, jelly-like, translucent, cream-coloured. This species, like other members of the genus, is a parasite of other fungi, particularly species of *Stereum*. *T. encephala* is mostly found in pine forest on decaying logs and branches. **Spores** c. 10 x 8 µm, subglobose to short ellipsoidal, smooth, colourless.

456

458

457

456 ***Tremella fimbriata*** [= *Tremella frondosa*]

Fruit bodies form irregular lobed colonies up to 80 mm across. The texture is very gelatinous. Occurs on dead wood in moist, shaded forest localities. **Spores** 10 x 9 µm, subglobose, smooth. Spore print white.

459

457 *Tremella fuciformis* Jelly Fungus.

Fruit bodies variable, forming small to large translucent, convoluted masses up to 160 mm long. This fungus is amongst the early species to appear on fallen tree trunks and branches and is common in eucalypt and mixed forest. **Spores** c. 8 x 8.5 µm, globose, smooth, colourless.

458 ***Tremella globispora*** [= *T. candida* var. *globospora*]

Fruit bodies variable, c. 15–25 mm across, irregularly rounded and uneven, translucent white. It is a parasitic crust-forming fungus that grows on twigs and branches of various native and introduced trees. It is distinguished from *T. encephala* (**455**) by its white colour and smoother surface. **Spores** c. 7.5 x 7 µm, globose to subglobose, smooth, colourless.

459 *Tremella mesenterica*

Fruit bodies variable in size. In dry forest it forms small compact colonies on dead twigs and branches. In moist to wet forest it forms much larger, brain-like masses on dead logs and branches. The colour is deep orange, particularly on banksia wood, becoming paler as the fruit bodies enlarge. **Spores** c. 14 x 8 µm, broadly ellipsoidal, smooth, colourless.

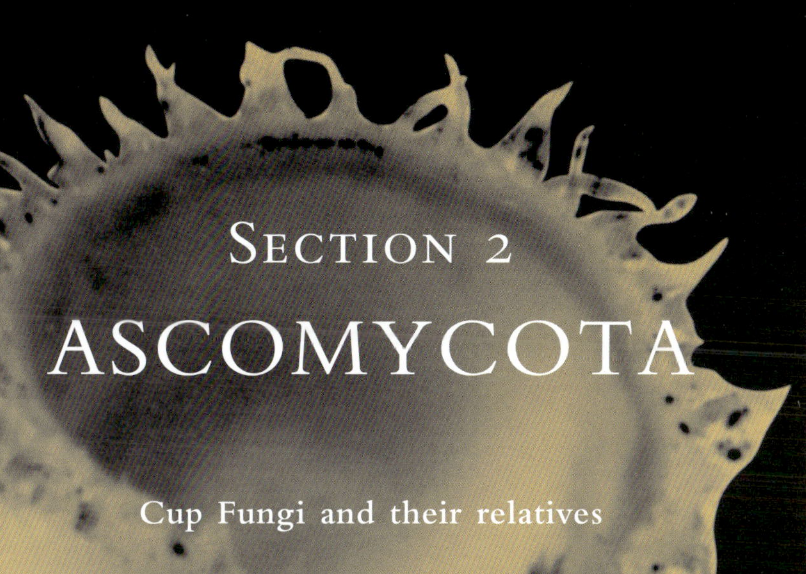

Section 2

ASCOMYCOTA

Cup Fungi and their relatives

The Ascomycota produce their spores in microscopic tubes or sacs called asci, that form the fertile layer. In members of this Division are found some diverse kinds of fruiting bodies. The cup fungi form a large group that contains species in which the fertile layer is cup-shaped or flat to convex. In other groups the fruiting body can be club-shaped, as in *Cordyceps*, or encrusted with a pitted surface, as in *Hypocrea*.

460 *Aleuria aurantia* Orange-peel Fungus

Fruit bodies to 70 mm across, at first concave becoming broadly concave to flattened and undulate. Gregarious to caespitose, often contorted by mutual pressure. This brightly coloured cup-fungus may easily be mistaken for discarded orange peel when seen along the roadside. Blowing sharply on mature fruit bodies will often release clouds of white spores (**460b**). **Spores** c. 20 x 10 µm, ellipsoidal, including net-like ornamentations and end extensions.

460a

461 *Aleuria rhenana*

Fruit bodies to 25 mm across, usually deeply concave with distinct, pale stalks. The individual cups usually remain concave in shape when mature. Gregarious to caespitose on earth. **Spores** c. 25 x 12 µm, ellipsoidal, including net-like ornamentations and end extensions, colourless.

462 *Aleurina ferruginea*

Discs to 20 mm across, concave at first, becoming saucer-shaped with a raised rim. The rusty brown outer surface is finely velvety, with scattered small brown warts. Scattered to densely gregarious. Widespread on the ground in various habitats. **Spores** c. 25 x 14 µm, ellipsoidal, ornamented with large tubercles. Spore print white.

463 *Anthracobia muelleri*

Discs to 3 mm across, slightly concave or flat with margins irregularly beset with short hairs. Gregarious to densely crowded and contorted. This fungus is always associated with burnt areas and frequently colonises large tracts of ground after bushfires. **Spores** c. 18 x 7 µm, oblong-ellipsoidal, smooth, colourless.

464 *Anthracobia* aff. *maurilabra.*

Discs to 8 mm across, occasionally larger. This *Anthracobia* is close to *A. maurilabra*, but is much darker in colour. It occurs after fire, often in white quartz sandy locations. **Spores** c. 20 x 8 µm, narrow-ellipsoidal, smooth, colourless.

465 *Ascocoryne sarcoides*

Discs to 20 mm across, single or sometimes confluent, with irregular darker margins. Texture gelatinous. The conidial stage (**465b**) is distinguished by pale conidia-covered clubs. Gregarious to caespitose on dead logs and branches in wet sclerophyll forest and rainforest. It seems to prefer *Acacia* species. **Spores** c. 14 x 5 µm, ellipsoidal with one septum or division, smooth, colourless.

465a

465b

Banksiamyces is an Australian ascomycete genus confined to *Banksia* cones. Most species are host-specific, but occasionally two species occur on one *Banksia* cone. Fruit bodies are discoid when moist but become laterally compressed with inrolled margins on drying. When rehydrated they resume their former discoid form.

466 *Banksiamyces katerinae*

Discs c. 3 mm across. It is usually found on the outer edge of the valves of cones of *Banksia ornata* and is probably the smallest species in the genus. **Spores** c. 7 x 3 µm, ellipsoidal, smooth, colourless.

467 *Banksiamyces maccannii*

Discs c. 4 mm across. It appears to be specific to *Banksia saxicola*. **Spores** c. 12 x 5 µm, ellipsoidal, smooth, colourless.

466

467

468 *Banksiamyces macrocarpus*

Discs 15 mm or more across, the largest species in the genus and known only from cones of *Banksia spinulosa*. The discs are prominent when moist but are nearly invisible when dry. **Spores** c. 6 x 2.5 µm, ellipsoidal, smooth, colourless.

469 *Banksiamyces toomansis*

Discs to 5 mm across, with a powdery appearance externally, usually occurring on cones of *Banksia marginata*, but occasionally found on other *Banksia* species. **Spores** c. 7 x 3 µm, ellipsoidal, smooth, colourless.

470 *Bisporella citrina*

Discs to 3 mm across, usually convex or flattened with age, almost stemless. A gregarious fungus forming small to large colonies on dead wood and bark. It is also common on dead cones of banksia and pine. Cosmopolitan. **Spores** c. 10 x 3.5 µm, elongate-ellipsoidal with one division, smooth, colourless.

471 *Bisporella oritis*

Discs to 1 mm across, but mostly smaller, deep yellow to orange. This tiny fungus is confined to the dead seed follicles of *Orites lancifolia*, an alpine shrub in the family Proteaceae. It is mostly found in summer before the old follicles are shed. **Spores** c. 8 x 3.5 µm, broadly ellipsoidal, smooth, colourless.

470

471

472 *Boedijnopeziza insititia* [= ***Microstoma insititium***]

Cups to 15 mm across, deeply concave to wineglass-shaped. Pale brownish buff with paler stem. The cap margin is beautifully ornamented by two or three rows of delicate broad-based scales. A rare tropical rainforest fungus found on decaying wood. **Spores** c. 45 x 9 µm, curved like a shallow boomerang, tapered at both ends, smooth, colourless.

473 ***Cheilymenia raripila***

→ Discs to 2 mm across, sparsely fringed with pale brown hairs. Gregarious on herbivore dung. **Spores** c. 25 x 12 µm, ellipsoidal, smooth, colourless.

474 *Chlorociboria aeruginascens* [= *Chlorosplenium aeruginascens*]

Discs variable c. 3–7 mm across with short, central or eccentric stem. Colour variable from bright to dark blue-green. The wood on which the fungus grows is stained blue-green and has been used in the manufacture of inlaid furniture known as Tunbridge Ware. Frequent on decaying wood in moist forest. **Spores** c. 8 x 2 µm, cylindrical with bluntly pointed ends, smooth, colourless.

475 *Chlorociboria* sp.

Discs to 3 mm across with short central stem. Grows on dead wood lying on moist ground. It resembles *C. aeruginascens* (474), but barely stains the wood on which it grows. The spores also differ in being much broader in the centre with a central septum or division. **Spores** c. 9 x 3.5 µm, spindle-shaped, with one division and rounded ends, smooth, colourless.

476 *Chlorovibrissea bicolor*

Fruit bodies to 18 mm high with fertile heads c. 5 mm across. A rare ascomycete of wet, shaded fern gullies and rainforest, where it grows on waterlogged and semi-submerged wood. **Spores** c. 60 x 2 µm, filiform, slightly tapering, smooth, colourless.

477 *Chlorovibrissea melanochlora*

Fruit bodies to 30 mm high, the fertile heads up to 7 mm diam. It usually grows in rows or clusters on dead wood, sometimes rising from cracks and cavities in decaying logs and branches in wet eucalypt forest and rainforest habitats. Uncommon. **Spores** up to 120 x 11.5 µm, filiform with cross walls, colourless.

478 *Cookeina tricholoma*

Cups to 20 mm diam., deeply concave and covered externally with bristle-like hairs, particularly on the cup margin (**478b**). The hairs are deciduous, and older specimens may lack them altogether. A tropical fungus found on dead wood in rainforest. **Spores** c. 28 x 12 µm, ellipsoidal, with longitudinal ridges, bluntly pointed at ends, colourless.

478a

478b

479a

479b

Cordyceps – Vegetable Caterpillars

Cordyceps is a genus of highly specialised fungi that parasitise insects, usually in their larval stage. Moth larvae of the genus *Oxycanus* are hosts for the most common species, *C. gunnii*, *C. hawkesii*, *C. robertsii* and *C cranstounii*, which all form large fruit bodies. The thread-like spores produced by the mature clubs break up into minute part-spores c. 3–5 mm long that penetrate the soil. They infect the caterpillars and consume the soft tissue. A column of fungal tissue then emerges, usually from behind the head of the dead caterpillar, terminating in the fertile clubs that emerge from the soil. Several smaller *Cordyceps* species are known, but not often seen.

479 *Cordyceps cranstounii*

This species produces a cluster of cream-coloured, club-shaped fruit bodies up to 20 mm high, covered with conspicuous ostioles. The tissue below ground connecting the caterpillar to the fertile heads is distinctively branched and lace-like (**479b**). Widespread in various forest habitats. **Part-spores** c. 3 x 1.5 µm, cylindrical, smooth, white.

480 *Cordyceps gunnii*

Fruit bodies to 100 mm high, elongated club-shaped with yellow stem grading into the dark olive-green fertile head (**480b**). Found in forest and usually associated with *Acacia* species. **Part-spores** c. 3.5 x 1.5 µm, cylindrical, smooth, white.

481

483

481 *Cordyceps hawkesii*

Fruit bodies to 70 mm high, the fertile head, with roughly cylindrical and rounded apex, distinctly differentiated from the stem. Widespread in eucalypt and mixed forest. **Part-spores** c. 4 x 2 µm, cylindrical, smooth, white.

482 *Cordyceps menesteridis*

Fruit bodies to 30 mm long, the fertile clubs ovoid, c. 5–6 mm long, dotted with small brown ostioles. This distinctive but uncommon species is found on beetle larvae. Occasional in moist forest. **Part-spores** not examined.

483 *Cordyceps robertsii*

Fruit bodies to 100 mm high, single or branched, terminating in long slender points. The fertile tissue is clearly distinct from the sterile stems. The peculiar white or cream outgrowths frequently seen on this fungus are *C. cranstounii* tissue, which appears to parasitise the *C. robertsii* host. Widespread in eucalypt forest. **Part-spores** c. 4 x 2 µm, cylindrical, smooth, white.

484 *Cudoniella pezizoidea* [= ***Hymenoscyphus sp.***]

Discs to 15 mm across, circular or variously compressed and convoluted when forming dense colonies. Colour from cream to pale greenish grey. Found on decaying forest debris in moist, shaded localities. **Spores** c. 26 x 5 µm, long-ellipsoidal to fusoid, curved, smooth, with oil bodies. Spore print white.

482

484

485 →

485 *Hispidula dicksoniae*

← Fruit bodies to 2 mm across, depressed, with distinctive, erect, delicate, broad-based scales around the margins of the tiny discs. A distinctive fungus apparently restricted to dead stems of tree-ferns in the genus *Dicksonia*. Rare. **Spores** c. 11.5 x 3 µm, ellipsoidal with cross wall, colourless.

486 *Cyttaria gunnii* Beech Orange

Fruit bodies to 25 mm across. In Australia it is restricted to myrtle beech, *Nothofagus cunninghamii*, on which it forms woody galls. The fruit bodies appear from late spring to summer, hanging in clusters from the galls. A membrane envelops the immature fruit body, which ruptures as the ball expands, revealing a pattern of concave depressions. **Spores**, 12 x 7–12 µm, subglobose. Spore print black.

486

487 *Daldinia concentrica* [= ***Daldinia grandis***]

Fruit bodies to c. 40 mm across, unevenly subspherical, firm, becoming brittle with age. At first brownish purple and dotted with fine ostioles, becoming black with age. Cut specimens show concentric growth zones. The perithecia form a layer beneath the surface crust. Found on dead wood in wet forests. **Spores** c. 14 x 17 µm, bean-shaped with lateral groove, smooth. Spore print black.

488 ***Discinella terrestris***

Discs to 10 mm across, but usually less. Discs are orange to yellow-orange and are flattened to undulate. It is found on the ground often amongst moss, and at the base of trees, where it grows on the lower regions of moss-covered trunks. **Spores** not measured. Spore print white.

489 *Fabraea rhytismoides*

A distinctive ascomycete found in colonies on the upper surface of the leaves of the montane daisy, *Leptinella filicula*, where it forms small, packed colonies with individual discs up to 1 mm across. **Spores** c. 18 x 5 µm, ellipsoidal-filiform with up to three cross walls, colourless.

490 *Galiella* aff. *celebica* [= ***Galiella aff. Rufa***]

Fruit bodies to 50 mm across, very thick and solid, tub-shaped with flat top. Finely hairy externally. A peculiar rainforest species photographed in Queensland near Kuranda. **Spores** not examined.

491 *Geoglossum* sp. Earth Tongues

Fruit bodies to 70 mm high, the fertile clubs may be cylindrical, grooved or flattened and are easily distinguished from the sterile stem. There are several species and they are usually found on the ground amongst moss. **Spores** c. 80 x 6 µm, cylindrical, slightly curved with up to eight divisions, smooth, olive-brown.

492 *Geopyxis carbonaria*

Cups variable, 7–10 mm across, sometimes larger, frequently forming extensive colonies on burnt ground after bush fires. It often occurs with other cup fungi with similar habitat preferences such as *Anthracobia* spp. and some *Peziza* spp. **Spores** c. 14 x 7 µm, ellipsoidal, smooth, colourless.

490

492

493 *Gyromitra esculenta* False Morel

Fruit bodies variable to 100 mm across, usually less, irregularly convoluted to brain-like, pale brown to dark red-brown. Stem simple to irregularly furrowed, white to brown. Single to gregarious, usually on sandy soil. Uncommon. **Spores** c. 20 x 10 µm, ellipsoidal, smooth. Spore print white.

494 *Helvella chinensis* [= *Helvella fibrosa*]

Fruit bodies to 25 mm across. The outer surfaces of the disc and stem are covered with short, soft hairs. An uncommon fungus usually associated with myrtle beech forest, sometimes found in wet eucalypt forests. **Spores** c. 18 x 12 µm, ellipsoidal, colourless; mature spores smooth, immature spores warty.

495 *Hyalopeziza* aff. *alni*

Fruit bodies to 1 mm across are found on dead branches and twigs, and are easily overlooked. The dense fringe of hairs around the disc is a distinctive character. Uncommon. **Spores** c. 10 x 2 µm, spindle-shaped, with one end broader, smooth, colourless.

496 *Hypocrea rufa*

Fruit bodies to 12 mm across, convex, lobed, with pale margins when young. Surface smooth, minutely dotted with ostioles. Found on dead wood in moist to wet forest habitats. **Part-spores** c. 4.5 x 4 µm, subglobose, minutely spotted, colourless.

493

494

495

496

497 *Hypocrea sulphurea* [= ***Hypocrea victoriensis***]

Fruit bodies form cushions or patches on dead wood in wet forests and rainforests. The ostioles are clearly visible on the surface. **Spores** c. 3.5 to 5 µm, subglobose, minutely spotted, colourless.

498 ***Hypocreopsis amplectens***

Fruit body to 40 mm across, consisting of a central region with irregular radiating lobes dotted with ostioles. Recorded from southern Victoria growing on coastal *Leptospermum*. Rare. **Spores** c. 30 x 9 µm, broadly spindle-shaped with one cross-wall.

499 *Hypoxylon bovei* [= ***Annulohypoxylon bovei***]

Fruit bodies form hard, black, brittle crusts on dead wood. The ostioles are surrounded by a black margin, giving each the appearance of a minute volcanic cone. Widespread in various forest habitats. **Spores** c. 12 x 7 µm, ellipsoidal with a groove, one side flattened, smooth, black.

500 ***Hypoxylon sp.***

Fruit bodies form brightly coloured crusts on the under side of decaying logs of myrtle beech (*Nothofagus cunninghamii*). Uncommon. **Spores** not examined.

501 *Hypoxylon fuscum* [= ***Hypoxylon howeianum***]

Fruit bodies to 6 mm across. Perithecia c. 0.3 mm across, embedded in an outer layer of tissue. Fruit bodies can be scattered, gregarious or in tightly adpressed colonies on bark or bare dead branches. Collected specimens soon become blackened with discharged spores. **Spores** c. 14 x 7 µm ellipsoidal with one flat side, smooth, dark brown.

502 *Hypoxylon rubiginosum* [= ***Hypoxylon placentiforme***]

Fruit bodies form thin, brittle, often extensive purplish crusts on bare, dead wood. The minute white dots on the surface are the openings of the pores through which the spores are discharged. The surface becomes black with age. Common. **Spores** c. 10 x 5 µm, ellipsoidal, flattened on one side with groove, black.

503 ***Inermisia fusispora***

Discs to 3 mm across, in gregarious to densely crowded, compressed colonies. Colour variable from creamy yellow through orange to almost red. Found on moist ground and plant debris, especially after fire. Uncommon. **Spores** c. 24 x 9 µm, fusoid with rounded ends, smooth. Spore print white.

502

503

504 *Lachnum pteridophyllum*

Discs minute, to c. 0.5 mm across, densely hairy externally. Gregarious on the stems of dead tree-fern fronds. Common in wet forest and rainforest where tree-ferns occur. **Spores** 20 x 5 µm, cylindric-fusoid, smooth. Spore print white.

505 *Lachnum virgineum*

Discs to 2 mm across, goblet-shaped, densely covered with short, white hairs. Frequent on dead wood and bark in moist situations in eucalypt forest and rainforest. **Spores** c. 10 x 2 µm, cylindric-fusoid, smooth.

506 *Lamprospora tuberculata*

Discs 1–3 mm across, more or less flat with a distinct membranous margin, scattered to gregarious amongst mosses of several species, especially *Campylopus spp.* and occasionally with liverworts. The genus name, *Lamprospora*, meaning 'beautiful spores' refers to the large globose spores, covered with tubercles which stain deep blue in Cotton Blue stain reagent. **Spores** c. 20 µm diameter, globose with prominent tubercles. Spore print white.

507 *Lanzia lanaripes*

Discs to 10 mm across with the tapering stem as long, or longer, than the disc diameter. The dark fruit bodies often have an olivaceous tinge and resemble broad-headed nails protruding from the rotting logs on which they grow. Found in wet forest and rainforest. **Spores** c. 15 x 5 µm, narrow-ellipsoidal to somewhat fusoid, but variable. Spore print white.

506

507

508 *Leotia lubrica*

Fruit bodies to 50 mm high with viscid stems. The fertile heads are roughly globose, or sometimes irregular. This cosmopolitan species grows singly or in small colonies amongst leaf litter in wet forest and rainforest. **Spores** c. 25 x 6 μm, long-ellipsoidal, curved, with up to 5 cross walls, colourless.

509 *Leucoscypha catharinaea*

Discs to 5 mm across. This specialised fungus grows only on moss, especially *Atrichum androgynum*, along shaded stream banks or in the spray of waterfalls. Uncommon. **Spores** c. 23 x 11 μm, ellipsoidal, with irregular warts, colourless.

508

509

510 *Microglossum viride*

Fruit bodies to c. 50 mm high, round, grooved or somewhat flattened and irregular in section. The sterile base is darker than the fertile head. It resembles a green *Geoglossum* but is distinct microscopically. Subtropical, also in Northern Hemisphere. Uncommon. **Spores** c. 17 x 5 μm, spindle-shaped, obscurely septate with faint banding, colourless.

511

511 *Mitrula* sp.

Fruit bodies to 15 mm high. The smooth heads are often flattened or grooved and supported by darker, sometimes slightly floccose stems. This species appears to be restricted to the fibrous trunks and bases of tree-ferns. **Spores** of illustrated species c. 12 x 3.5 μm, cylindrical with one cross wall, colourless.

512 *Morchella conica*

Fruit bodies to 140 mm high, rather variable in size and colour. Usually found after fire and in similar habitats to *M. elata*, but much less common. **Spores** c. 22 x 12 μm, broadly ellipsoidal, smooth. Spore print cream-coloured.

513 *Morchella elata* Morel

Fruit bodies to 120 mm or more high. Found in open forest, including mallee, often fruiting freely, usually in spring and after fire. Edible. Widespread but uncommon. **Spores** c. 22 x 14 µm, broadly ellipsoidal, smooth. Spore print cream-coloured.

512

513

514

514 *Neobulgaria pura*

← Discs usually to 10 mm across but can grow much larger, forming crowded colonies on decaying wood and tree-fern trunks. The discs are very gelatinous and rubbery, becoming almost invisible when dry, expanding again when moistened. **Spores** c. 7.5 x 4 μm ellipsoidal, smooth. Spore mass white.

515 *Paecilomyces tenuipes* [= ***Cordyceps takaomontana***]

Fruit bodies to 50 mm high, the branches covered with powdery conidia. This uncommon but distinctive fungus seems to selectively parasitise beetle larvae. Widespread in rainforest gullies but uncommon. **Spores**/conidia not examined.

515

516 *Paxina costifera*

Cups to 40 mm or more across with distinct ridges on the external basal region. They become flattened or saucer-like with age. Found on lime-rich soils, often with pines. **Spores** c. 20 x 12 µm, broadly ellipsoidal, smooth. Spore mass white.

517 *Peziza austrogeaster*

Fruit bodies to 40 mm across; they develop underground, splitting into lobes which open at ground level. It resembles an earth star without the central puffball. It usually occurs in sand or sandy soils in arid and semi-arid regions. Common in mallee habitats. **Spores** c. 25 x 15 µm, ellipsoidal with irregular end ornamentation, smooth. Spore mass white.

518 *Peziza echinospora*

Cups to 80 mm or more across, frequently in clusters and often distorted by mutual compression. It is usually found in damp places where wood has been burnt. Flesh very brittle. **Spores** c. 15 x 7 µm, ellipsoidal, finely warted. Spore mass white.

519 *Peziza repanda*

Cups to 120 mm or more across. This very large cup fungus is found on a variety of organic materials, including dead wood and discarded household waste such as carpets, paper and rotting vegetation. The specimen shown was growing on a dead apple tree. **Spores** c. 14 x 9 µm, ellipsoidal, smooth. Spore mass white.

520 *Peziza tenacella*

Cups to 30 mm across, translucent-violet when young, gradually turning brown with age. Gregarious to caespitose on burnt ground. Often common after bush fires. **Spores** c. 11 x 6 µm, ellipsoidal, minutely rough. Spore mass white.

521 *Peziza thozetii*

Cups to 25 mm across, brown to olive-brown with inrolled margins, especially when immature. Usually found on lime-rich soils in pine plantations, otherwise uncommon. **Spores** c. 25 x 10 µm, ellipsoidal, finely warty with conical ornamentation at each end, pale brown.

522

523a

523b

524

522 *Peziza vesiculosa*

Fruit bodies to 80 mm across, at first hemispherical, becoming deeply cup- or urn-shaped, usually with inrolled margins. This large gregarious cup fungus is usually found on manure, old straw or decaying wood mulch. **Spores** c. 20 x 10 µm, ellipsoidal, smooth, colourless.

523 *Hydnoplicata convoluta* [= *Peziza whitei*]

Fruit bodies to c. 40 mm across, irregularly subglobose with folds and convolutions, frequently half-buried at maturity. Cut specimens (523b) show a complex arrangement of irregular chambers lined with fertile hymenium. A distinctive species, widely distributed but uncommon. **Spores** c. 10 x 12 µm, ellipsoidal, with a network of ridges, colourless.

524 *Plectania campylospora*

Cups to 50 mm across, the base contracting into a grooved stem. A large cup fungus of rainforest, fern gullies and moist, shaded slopes and gullies of drier forests. Gregarious to caespitose on decaying wood. **Spores** up to 30 x 12 µm, jellybean-shaped, smooth, colourless.

525 *Poronia Erici*

→ Fruit bodies to 10 mm across, irregularly round with a finely wrinkled surface scattered with large ostioles. It occurs on herbivore dung including kangaroo, wallaby, wombat, rabbit and occasionally horse dung. **Spores** c. 25 x 16 µm, ovoid to bean-shaped, smooth, dark brown.

526 *Pyronema omphalodes*

The tiny fruit bodies, c. 0.5 to 1 mm across, are compressed into irregular patches with a pink to white mycelial margin around growing colonies. Often the first fungus to appear after fire, sometimes covering large areas. Also common on campfire sites. **Spores** c. 16 x 10 µm, ellipsoidal, smooth. Spore mass white.

527 *Sarcoscypha coccinea*

Cups to 40 mm across with a short stem. It resembles *Aleuria rhenana* but is larger, scarlet and has smooth spores. Usually on the forest floor amongst mosses and leaf litter. Uncommon. **Spores** variable, from c. 20–40 x c. 12 µm, ellipsoidal, smooth, white.

528 *Scutellinia scutellata*

Discs to 10 mm or more across, gregarious, often deformed by crowding. Margin of disc bearing radiating stiff dark hairs. Found on wet ground and waterlogged decaying wood and bark in shaded habitats. **Spores** c. 18 x 12 µm, ellipsoidal, minutely ornamented. Spore mass white.

527

528

529 *Stictis radiata*

Discs to 1 mm across but usually less. The tiny discs have a white fluffy margin. The colourless spores are relatively enormous by comparison to the size of the disc. Gregarious on dead wood and forest debris in moist to wet forest. **Spores** up to 200 x 5–7 µm, colourless.

530 *Torrendiella madsenii*

Discs to 10 mm across with a fringe of short hairs on the margin. Discs at first pale blue-grey drying yellow or cream-coloured. Found on decaying wood in moist to wet forests. Gregarious or in small clusters. **Spores** c. 18 x 4 µm, spindle-shaped, colourless.

529

530

531 *Trichoglossum hirsutum*

Fruit bodies to 50 mm high, closely resembling the related genus *Geoglossum*. The stems and fertile heads of *Trichoglossum* are covered with dark bristles. Heads can be irregularly subglobose, ellipsoidal, or club-shaped and are often compressed or grooved. Found on the ground, usually amongst moss. **Spores** c. 150 x 6 µm, cylindrical with up to 15 partitions, smooth, dark green.

532 *Underwoodia beatonii*

→ Fruit bodies to 100 mm high, single or gregarious, resembling undeveloped or malformed morels. Cut specimens show long hollow chambers lined with spore-bearing tissue. It is associated with several species of *Melaleuca* in mallee and coastal regions. **Spores** c. 24 x 11 µm, ellipsoidal, ornamented with irregular warts. Spore print white.

533 *Vibrissea dura*

Fruit bodies up to 30 mm high, the fertile heads to 8 mm across. Colour may vary from cream-coloured heads with white stalks to pale brown all over. Found on decaying logs and fallen branches in wet forest and rainforest. Uncommon. **Spores** c. 120 x 2 µm, filiform, colourless.

534 *Xylaria hypoxylon*

Fruit bodies to c. 80 mm high, roughly cylindrical, club-shaped or flattened, frequently branched. Covered with white powdery conidia when young, black when mature. Stems hairy. Common on dead wood in moist forests. **Spores** c. 12 x 5.5 µm, bean-shaped, smooth. Spore print black.

535 *Xylaria polymorpha* [= *Xylaria castorea*] Dead Man's Fingers

Fruit bodies to c. 80 mm high, irregularly cylindrical to club-shaped, often grooved or flattened. The black clubs are covered with fine pimple-like ostioles. When cut, the white flesh highlights the dark perithecia embedded below the surface. Colonises dead wood in wet forests. **Spores** c. 10 x 5.5 µm, nearly bean-shaped with one flattened side, blackish brown.

535

SECTION 3
MYXOMYCOTA

Slime Moulds

The Myxomycota, commonly known as slime moulds, are now classified in a separate kingdom, Protoctista. They are a distinctive group of organisms sharing some of the characters of amoeba and fungi. The mature fruit body, called the sporangium (538), produces spores that eventually develop into a creeping, slime-like mass called a plasmodium (537). The plasmodium behaves like a giant amoeba, feeding on bacteria, fungi and decaying organic matter. When the food supply is exhausted, or in response to changing conditions, the plasmodium undergoes a transformation to once again produce a sporangium.
The slime moulds are often included in the fungal literature and are introduced here to stimulate interest in these relatively little known organisms.

536 *Ceratiomyxa fruticulosa*

Sporangia on rotting wood. **Spores** c. 11 x 6.5 µm, ellipsoidal, smooth.

537 *Dictydiaethalium plumbeum*

Intermediate phase; plasmodium, on wood chips, changing to produce sporangia. **Spores** c. 10 µm diam., spheroid, spiny, brown.

538 *Fuligo septica*

Sporangium-producing phase on leaf litter. **Spores** 6–8 µm diam., globose, finely spotted, purplish black

539 *Fuligo sp.*

Sporangium on ground and leaf litter in mallee scrub.

540 *Lycogala epidendrum*

Sporangium-producing phase on dead wood. Fruit body with broken surface shows transition stage. **Spores** 4–7 µm diam., globose, reticulate, brown.

541 *Stemonitis axifera*

Spores c. 8 µm, globose, finely warted, brown. *Stemonitis* sp. (**541b**) changing to produce sporangia.

542 Slime mould (unidentified) on moss.

542a Plasmodium **542b** Sporangia

541b

542a

542b

543a

543b

545

543 Slime mould (unidentified) on leaf litter.

543a Plasmodium 543b Sporangia.

544 Slime mould (unidentified) on *Eucalyptus obliqua bark.*

544a Plasmodium 544b Sporangia.

545 *Diderma* sp.

Sporangia on dead *Xanthorrhoea* flower spike.

546 Slime mould (unidentified)

546a Plasmodium 546b Sporangia-producing phase.

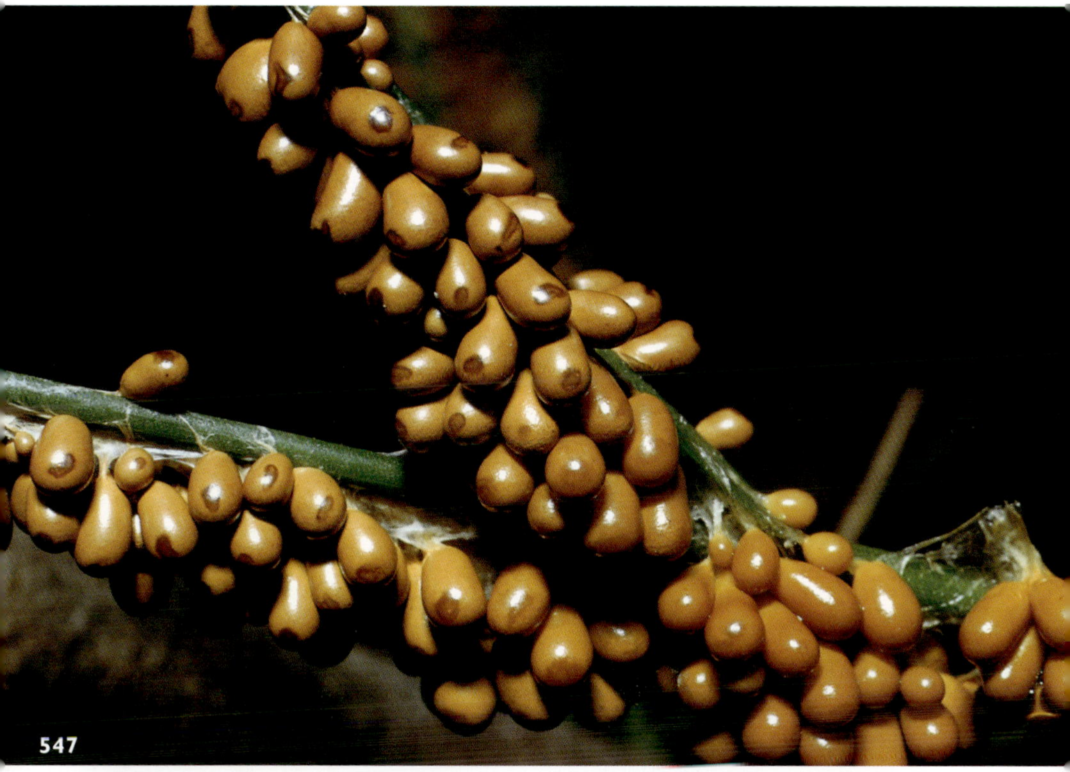
547

548

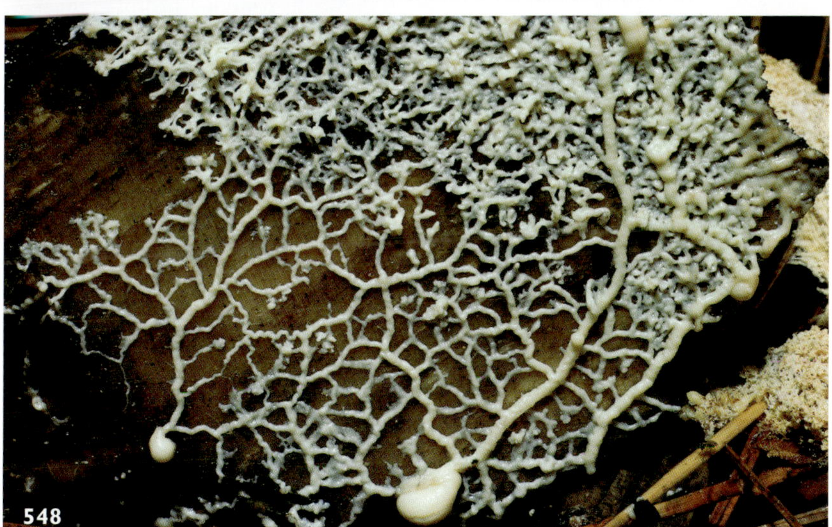

547 *Leocarpus fragilis*

Sporangia fruiting on grass.

548 Slime mould (unidentified)

Plasmodium in active mobile phase.

Glossary

aff. with affinities to.

amyloid blue colour reaction of fungal spores or hyphae to iodine (Melzer's reaction).

annulus a collapsed veil forming a ring of tissue around the stem of some mushrooms.

ascus (pl. **asci**) tubular or sac-like cells containing spores of ascomycetes.

basidium (pl. **basidia**) the spore-producing organ of a basidiomycete fungus.

c. (circa) about.

caespitose several fruit bodies arising from a common base (see also **gregarious).**

capitate forming a rounded head.

chlamydospores thick-walled cells of swollen hyphae.

confluent running into one another (e.g. as gills with fruit body stem).

conidia (adj. **conidial**) asexual spores formed from hyphae.

cortina a cobweb-like veil between cup and stem of some mushrooms.

cystidium (pl. **cystidia**) sterile cells projecting from gills and other parts of a mushroom.

deciduous shed or falling off at the end of its growth period.

decurrent (*in gills*) continuing down the stem for some distance.

diam. diameter.

ellipsoidal elliptical in section.

exoperidium the outer skin or layer of a puffball.

fertile tissue the layer of tissue from which spores are produced.

fibril (adj. **fibrillose**) a fine hair.

filiform thread-like.

flabelliform in the shape of a fan.

flexuose bending or curving in a wavy or zig-zag manner.

floccose with soft woolly or granular covering which rubs off easily.

follicle a dry fruit formed from one carpel splitting along a single suture.

fruit body the large visible part of a fungus, i.e. a mushroom, toadstool, etc.

fugacious withering or falling off rapidly.

fusiform spindle-shaped, tapering at both ends.

gall woody outgrowth produced on host plants by certain fungi.

germ pore a modified, thin-walled region at one end of a spore.

gleba the spore-bearing tissue mass in puffballs and their relatives.

globose spherical, or nearly so, in shape.

glutinous slimy, very sticky.

granular (*of a surface*) covered with very small, rounded protuberances.

gregarious growing in troops or open clusters but not clumped or joined at the base (see **caespitose**).

hispid covered with stiff, bristly hairs.

hygrophanous changing on drying from translucent to opaque.

hymenium the spore-bearing layer of a fruit body.

hypha (pl. **hyphae**) thread-like cells of the mycelium. Initial product of spore germination

hypogeous developing or fruiting underground.

indusium (in *Phallus*) a veil that hangs from the top of the stem under the cap.

lacuna (adj. **lacunose**) a hole or cavity in the middle of tissue.

latex clear or coloured juice that bleeds from cut or bruised surfaces of some fruit bodies.

lenticular in the shape of a lens, i.e. with one or both surfaces convex.

lichenised associated with algal cells as in lichens.

luminescent emitting visible light.

metuloid (adj. **metuloidal**) large, thick-walled cystidium.

mycelium the vegetative structure of a fungus composed of a series of filamentous hyphae.

mycorrhiza (adj. **mycorrhizal**) fungus forming a symbiotic relationship with the root of a plant.

nitric nitric oxide odour of some fungi.

ostiole an opening or pore through which spores are discharged.

partial veil the protective membrane connecting the cap margin to the stem of an immature fruit body.

pathogen an organism capable of causing disease in a host.

pellicle a protective skin or film-like covering that usually removes easily.

peridiole a small division of the gleba having a separate wall.

peridium the limiting wall of the fruit body.

perithecium (pl. **perithecia**) an embedded flask of asci.

plasmodium the mobile feeding stage of a slime mould. A slime-like multinucleate mass of protoplasm.

pore any small opening. In fungi, the opening of the tubes in polypores and boletes.

pseudosclerotia (false sclerotium) masses of soil particles bound together by fungal hyphae.

punctate minute dots or depressions.

rachis the axis of a pinnate leaf; e.g. in many ferns.

rehydrate reabsorption of water by a dried fruit body.

rhizomorph root or cord-like aggregation of mycelium.

sclerophyll having hard leaves, like species of *Eucalyptus*.

sclerotium (pl. **sclerotia**) a compacted mass of mycelium tissue produced by some fungi as a food reserve.

sp. (pl. **spp.**) species.

spathulate spoon-shaped.

sporangium fruit body.

spore the microscopic reproductive units of fungi.

spore print spore deposit from fruit body.

squamule (adj. **squamulose)** a small, soft scale.

stellate star-shaped.

stoma (Greek., a mouth) an opening or pore.

striate of a surface when marked with fine lines.

sulcate of a surface when grooved or furrowed.

thallus (adj. **thallose**) a simple plant body not differentiated into roots, stem and leaves.

tomentum (adj. **tomentose**) down or fine matted hairs.

translucent-striate gill pattern visible through translucent tissue.

tubercle (adj. **tuberculate**) a small swelling, wart or rounded protuberance.

type specimen the specimen cited in the original description of the species.

type locality the place from which the type specimen was collected.

µm micron, one millionth of a metre.

umbo (adj. **umbonate**) a conical or rounded protuberance.

universal veil a protective membrane that envelops developing fruit bodies of some fungi.

viscid sticky.

volva swollen stem base with sac or collar remains of a universal veil.

Further Reading

Aberdeen, J.E.C. (1979), *An Introduction to Mushrooms, Toadstools and Larger Fungi of Queensland*. Queensland Naturalists' Club, Brisbane.

Bougher, N.L. & Syme, K. (1998), *Fungi of Southern Australia*. University of Western Australia Press, Nedlands, W.A.

Cleland, J.B. (1976), *Toadstools and Mushrooms and Other Larger Fungi of South Australia*, Parts I and II. South Australian Government Printer, Adelaide.

Cole M., Fuhrer B., Holland A. (1984), *A Field Guide to the Common Genera of Gilled Fungi in Australia*. Inkata Press, Hawthorn.

Fuhrer, B.A. (2001), *A Field Companion to Australian Fungi*, revised edn. Bloomings Books, Melbourne.

Fuhrer, B.A. & Robinson, R. (1992), *Rainforest fungi of Tasmania and south-east Australia*. CSIRO Press, East Melbourne.

Griffiths, K. (1985), *A Field Guide to the Larger Fungi of the Darling Scarp and South West of Western Australia*. Published by the author, Perth.

Grgurinovic, C.A. (1997), *Larger Fungi of South Australia*. Botanic Gardens of Adelaide, Flora & Fauna of South Australia Handbooks Committee, Adelaide.

Grgurinovic, C.A. (2003), *The Genus Mycena in south-eastern Australia*. Fungal Diversity Press, Hong Kong and ABRS, Melbourne and Canberra.

Grgurinovic, C.A. & Cayzer, Lindy Eds (2003), *Fungi of Australia Vol. 2B, Catalogue and Bibliography of Australian Fungi 2. Basidiomycota, Myxomycota*. ABRS/CSIRO Publishing, Melbourne.

Hall, I.R., Stephenson, S.L., Buchanan, P.K., Wung Yun, Cole, A.L.J. (2003), *Edible and Poisonous Mushrooms of the World*. New Zealand Institute for Crop and Food Research Ltd, Christchurch, New Zealand.

McCann, I.R. (2003), *Australian Fungi Illustrated*. Macdown Publications, Vermont, Victoria.

MacDonald, R. & Westerman, J. (1979), *A Field Guide to Fungi of South-Eastern Australia*. Thomas Nelson, Melbourne.

Negus, P. (2006), *The Magical World of Fungi*. Cape to Cape Publishing, North Fremantle.

Orchard, A.E. Exec. Editor (1996), *Fungi of Australia Vol. 1A, Introduction—Classification*, ABRS, Canberra.

— (1996), *Fungi of Australia* Vol. 1B, *Introduction—Fungi in the Environment*. ABRS, Canberra.

— (1997), *Fungi of Australia Vol. 2A, Catalogue & Bibliography of Australian Macrofungi 1. Basidiomycota*. ABRS Canberra

Peterson, R.H. (1988), *The Clavarioid Fungi of New Zealand*. DSIR Science Information Publishing Centre, Wellington, New Zealand.

Robinson, R. (2003) *Fungi of the South-West Forests*. Bush Books. C.A.L.M., Western Australia.

Stephenson, S.L. & Stempen, H. (1994), *A Handbook of Slime Molds*. Timber Press, Portland, Oregon.

Willis, J.H. (1963), *Victorian Toadstools and Mushroom*, 3rd edn. The Field Naturalists Club of Victoria, Melbourne.

Wood, A. (1990), *Australian Mushrooms and Toadstools. How to Identify Them*. New South Wales University Press, Kensington.

Young A.M. (2004), *A Field Guide to the Fungi of Australia*. University of New South Wales Press, Sydney.

Young, T. (1982), *Common Australian Fungi*. New South Wales University Press, Kensington.

Relevant web sites (current August 2004)

http://www.rbg.vic.gov.au/fungi/

http://www.deh.gov.au/biodiversity/abrs/publications/fungi/kingdoms.html

Schedule of Recent Name Changes

The name changes listed here derive from recent taxonomic revisions and DNA research on several fungus groups.

Page	Plate No.
35	28 *Tetrapyrgos olivaceonigra*
43	44 *Coprinopsis atramentarius*
50	58 *Cortinarius austroviolaceus*
56	67a & b *Cortinarius* sp.
56	56 *Cyptotrama asprata*
58	70 *Cystolepiota* aff.*sistrata*
58	71 *Cystolepiota* aff.*adulterina*
59	72 *Cortinarius austrovenetus*
60	73 *Cortinarius canarius*
60	74 *Cortinarius clelandii*
61	76 *Cortinarius kulus*
62	77 *Cortinarius persplendidus*
63	78 *Cortinarius magentiannulata*
71	92 *Galerina marginata*
71	93 *Galerina patagonica*
72	94 *Loreleia postii*
79	107 *Hygrocybe aurantiopallens*
80	109 *Bertrandia astatogala*
82	111 *Cantharellus lilacinus*
85	117 *Gliophorus graminicolor*
87	119 *Porpolomopsis lewellinae*
88	123 *Hygrocybe acutoconica*
91	125 *Gliophorus perplexus*
92	129a *Cuphophyllus virgineus*
92	129b *Cuphophyllus virgineus* var. *fuscesens*
95	135 *Leratiomyces ceres*
97	138 *Hypholoma australe*
98	139 *Hypholoma fasciculare* var. *americana*
107	154 *Lentinellus tasmanicus*
107	155 *Lentinellus castoreus*
108	156 *Lentinellus pulvinulus*
112	163 *Leucoagaricus leucothites*
118	172 *Leucocybe connata*
118	175 *Macrolepiota clelandii*
121	178 *Gymnopus foetidus*
124	185 *Deconica horizontalis*
129	191 *Roridomyces austrororidus*
143	213 *Cruentomycena viscidocruenta*
145	217 *Clitocybula* sp.
148	222 *Lichenomphalia chromacea*
149	223 *Lichenomphalia umbellifera*
150	224 *Lichenomphalia* sp.
153	229 *Panellus pusillus*
159	239a *Scytinotus longinquus*
167	251 *Rickenella swartzii*
179	272a & b *Volvopluteus gloiocephalus*
180	274 *Oudemansiella gigaspora*
185	281 *Pseudomerulius curtisii*
200	307 *Artomyces austropiperatus*
211	325 *Ramaria filicicola*
215	332 a&b *Lycoperdon pratense*
219	338 *Pisolithus albus*
235	362 *Phallus multicolor*
243	375 *Phlebia subceracea*
253	390 *Ganoderma australe*
253	391 *Ganoderma weberianum*
257	398 *Trametes apiaria*
259	400 *Osmoporus vesparius*
275	426 *Ceriporiopsiis merulinus*
288	447 *Pseudocraterellus undulatus*
290	450 *Calocera sinensis*
293	454 *Leucogloea compressa*
305	472 *Microstoma insititium*
312	484 *Hymenoscyphus* sp.
315	487 *Daldinia grandis*
316	490 *Galiella* aff. *Rufa*
318	494 *Helvella fibrosa*
320	497 *Hypocrea victoriensis*
321	499 *Annulohypoxylon bovei*
322	501 *Hypoxylon howeianum*
322	502 *Hypoxylon placentiforme*
330	515 *Cordyceps takaomontana*
341	535 *Xylaria castorea*

Index

➡*Accepted scientific names are in italics.*
➡Common names, synonyms and names misapplied to Australian taxa are not in italics.

Spore types

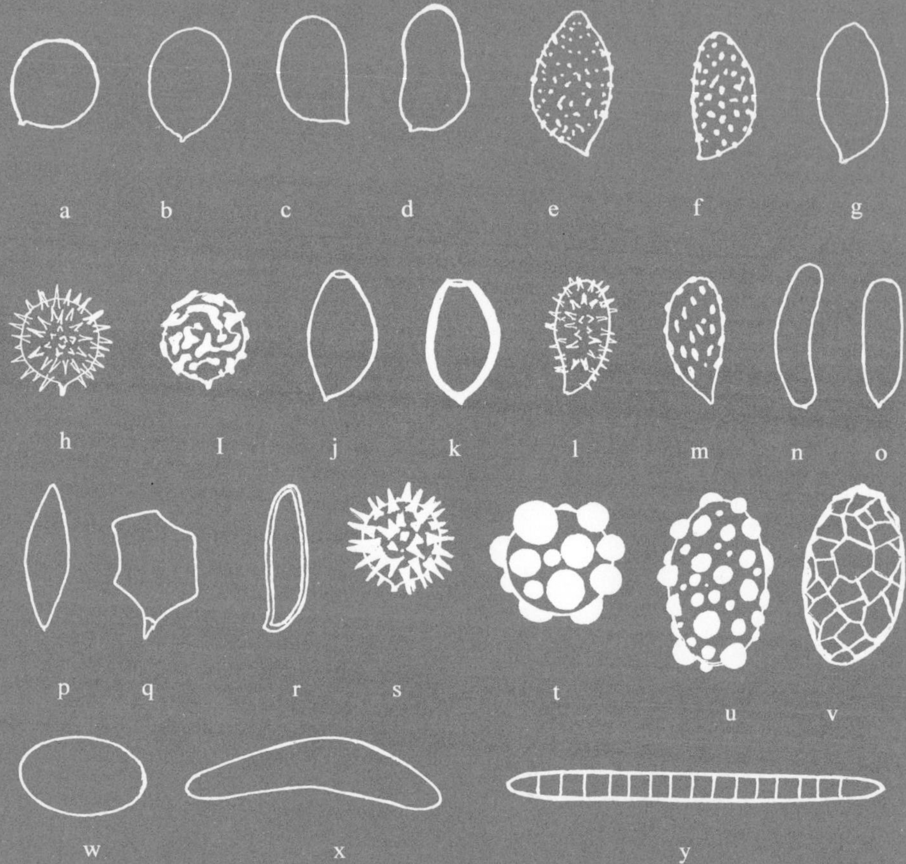

Examples of spore shapes viewed in profile

(a) globose (b) subglobose (c) ellipsoidal (d) ellipsoidal-constricted (e) lemon-shaped, warty (f) almond-shaped, warty (g) almond-shaped (h) globose, spiny (i) globose with amyloid warts and ridges (j) ellipsoid with germ pore (k) ellipsoid, thick walled with germ pore (l) tear-drop, spiny (m) tear-drop, warty (n) sausage shaped (allantoid) (o) cylindrical (p) spindle-shaped (fusoid) (q) angular (r) elongate, spindle-shaped (s) globose with dark spines (t) globose, tuberculate (u) ellipsoid, tuberculate (v) ellipsoid, reticulate (w) broadly ellipsoid (x) curved spindle-shaped (y) cylindrical, septate.